D1104968

SPACE GARBAGE

SPACE GARBAGE

Comets, Meteors and other Solar-System Debris
Guest Star: Halley's Comet

JACK MEADOWS

GEORGE PHILIP

Illustration Acknowledgments

Text illustrations

By courtesy of the British Museum (Natural History) 7,
  8, 23, 32,
R. L. Fleischer, P. B. Price and R. M. Walker 24
Mount Wilson and Las Campanas Observatories, Carnegie Institution of
  Washington 36 (pp. 114–15) (NASA), 43 (NASA), 44 (NASA),
  45, 46 (Science Photo Library)
IRAS Project 19
S. M. Larson and Z. Sekanina 40
Mansell Collection 13, 16, 33 (left), 39
NASA 3, 6, 15, 18, 21, 26, 27, 31, 41
Private Eye 34
Royal Astronomical Society 5, 14, 30, 36
Smithsonian Astrophysical Observatory 9, 10
Syndication International 11, 47
US Department of the Interior Geological Survey 25
US Naval Research Laboratory 20

Jacket illustrations

Stephen J. Edberg/International Halley Watch *front centre*
Royal Observatory, Edinburgh *front top*
Science Photo Library *front below left and below right, back*

British Library Cataloguing in Publication Data

Meadows, A. J.
  Space garbage: comets, meteors and other
  solar-system debris.
  1. Meteors 2. Meteorites. 3. Comets
  I. Title
  523.5'1   QB741
  ISBN 0-540-01087-1

© Jack Meadows 1985

First published by George Philip,
12–14 Long Acre, London WC2E 9LP

Printed and bound in Great Britain by
Biddles Ltd, Guildford and King's Lynn

Contents

Foreword

Astronomers can be divided into two groups – the Lords of Creation, who survey the whole Universe, and the pickers-up-of-unconsidered-trifles, who potter around in the solar system. Somewhere in the basement of the latter group come those astronomers whose special interest lies in the small lumps of matter that float about between the planets. Because this stuff is basically the left-overs from the bigger bodies, I have labelled it *space garbage*. However, its importance to astronomy is not to be measured in terms of its size. Archaeologists have acquired much of their knowledge of early human beings by sifting through the rubbish dumps they left behind. Similarly, much of what we know about the origin and early history of the solar system has been gained from a study of space garbage.

In this book I am trying to lay out, in the simplest terms I can manage, what space garbage is, and what it can contribute to astronomical knowledge. Since I have been in this field on and off for twenty years, I have described things, in part, from my own experience. I think I have covered all the main points that should be here, but I have also put in some of my personal viewpoints. These have not originated purely from my own investigations. I owe an immense debt to the research staff and students who have worked with me. I was supposed to be teaching them, but they generally taught me. I hope, if they ever get round to reading this book, they will recognize many of the topics we have discussed together in the past.

Jack Meadows

I
What is Space Garbage?

Imagine that you live on a planet which circles another star not too far from the Sun. Go out at night and look at the sky. You should be able to see the Sun as a moderately bright star. But it will seem to be alone in space: none of the planets around it will be visible. To see them you will need to take a lengthy space trip (several years, even if you travel by the fastest imaginable shuttle) to visit the Sun. When most of your journey is behind you, some of the planets will come into view; but not all will appear simultaneously. First of all, you will notice a group of large planets a fair distance away from the Sun. There are four of them – Jupiter, Saturn, Uranus and Neptune (in increasing order of distance from the Sun). You will have to approach closer still before you can pick up another group of planets, much smaller in size. These – again in order of distance from the Sun – include Mercury, Venus, the Earth and its attendant Moon, and Mars. You may also catch a glimpse of Pluto with its moon out towards the edge of the planetary system. The objects featured in this book – small bodies, maybe only a mile or two (a few kilometres) across, and the thin veil of gas and dust between the planets – can only be seen easily when you have actually entered the solar system.

Names are important in astronomy. The Universe contains so many different and improbable objects that it is vital to know which you are talking about. Many astronomical jokes depend on this need to be fussy about names. The classical examination question in astronomy is: 'Define the Universe, and give *two* examples.' So at intervals we will have to divert to discuss names. I have used two different names in the preceding paragraph – *planetary system* and *solar system*. The former simply means all the planets; the latter is a much more general term and covers everything within the Sun's sphere of influence (including the Sun itself). You might reasonably object that this is answering one question with another – what is meant by the Sun's *sphere of influence*? In astronomy most questions sooner or later get back to gravitation. The Sun is a very massive body, so it exerts a large pull – a gravitational pull – on any body near it in space. This is why planets and other bodies circle the Sun and do not move off into space. In theory, the Sun's gravitational pull does not disappear however far away you go (though its amount

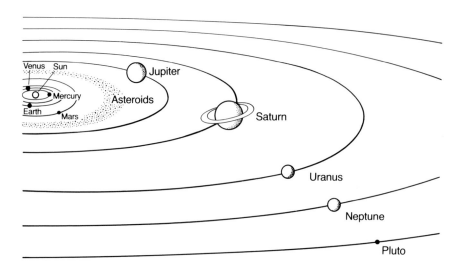

1 The important feature here is the great distances between the paths of the major planets compared with the small space in which the terrestrial planets orbit. Taking Figure 2 together with this one makes it obvious that an external observer would probably overlook the terrestrial planets in the glare of the Sun.

diminishes rapidly). In practice, once your space voyage has brought you half way to the Sun from the nearest star, the gravitational pull of the Sun will take over. So the Sun's sphere of influence can be thought of as a sort of imaginary balloon round the Sun, that goes out half way to the nearest stars. Anything that comes within this balloon is gravitationally attracted by the Sun, and therefore counts as part of the solar system (though not necessarily a permanent part).

The successive appearance of the different bodies as we approach the Sun corresponds quite well with the way we classify things in the solar system. The Sun is supreme. It provides the basic control on all motions in the solar system, as well as heat and light. For this reason, one method of classification looks at the bodies in terms of their orbits (that is the paths they follow) round the Sun. The simplest approach is to use the average distance of the different objects from the Sun. (The word 'average' comes in here because few bodies in the solar system move round the Sun in a perfect circle. Most move so that they are sometimes slightly closer to the Sun, and sometimes further away.) Distances in the solar system can be measured in any terms you like. You could say, for example, that the Earth ploughs its way round the Sun at an average distance of 93,000,000 miles (150,000,000 km). But it is not very convenient to handle numbers as large as this all the time. Astronomers therefore usually measure distances in the solar system in terms of the Earth's average distance from the Sun. Neptune, for example, circles the Sun at an average distance about thirty times greater than the Earth's distance from the Sun. The Sun–Earth distance is called an *Astronomical Unit* (often abbreviated to AU). An equivalent statement about Neptune would be that it is at 30 AU from the Sun.

The first group of planets you saw on your imaginary trip to the Sun included Neptune. This group is labelled the *major planets*. They became visible first because they are large, and so reflect a fair amount of sunlight. The nearest of these planets to the Sun is Jupiter at 5.2 AU (i.e. at just over five times the distance of the Earth). If eleven Earths were placed side by side across the disc of Jupiter there would still be some of the planet left peeping out round either end of the chain. This makes Jupiter the largest planet in the solar system; so it is often considered to be the prototype major planet. Consequently, the major planets are sometimes also referred to as the 'Jovian' planets.

You will recall that the next group of planets that came into view were much closer to the Sun (as the comparative distances of the Earth and Neptune mentioned above make evident). This group came into view later because they are all much smaller in size than the major planets. The Earth is the largest member, and so is often taken as the prototype. For this reason, the group is referred to collectively as the *terrestrial planets*. Whereas it is fairly easy to define a major planet, terrestrial planets present more problems.

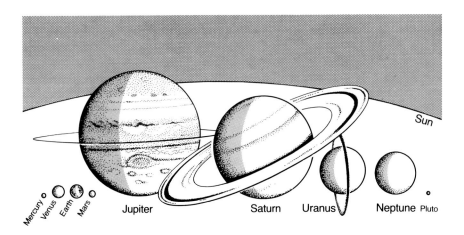

2 *The difference in size between the giant planets, like Jupiter, and the small planets, such as the Earth, is very striking. Less obvious is the fact that the latter group consist mainly of rock, whereas the former are vast balls of compressed gas. Note that at least three of the major planets are circled by belts of small particles – rather like small-scale models of the asteroid belt circling the Sun.*

The first problem relates to the way astronomers define planets. A *planet* is any relatively large body which follows its own independent path round the Sun. This distinguishes planets from moons (more generally called *satellites*), which are the smaller bodies that circle round planets, rather than moving directly round the Sun. For example, Mars has two small moons, only 10 to 20 miles across (15 to 30 km), which circle round it. The problem is that our own Moon is much larger than a satellite of the Earth should be. It is actually not very much smaller than the planet Mercury. Neither Mercury nor Venus has any satellites at all, so the Earth and Moon form a unique combination among the terrestrial planets. Some astronomers go so far as to call the Earth-Moon combination a double planet, and classify the Moon as a fifth terrestrial planet. Though most still label it a satellite, comparisons of the properties of the different terrestrial planets usually include the Moon in the discussion.

A second problem with the terrestrial planets is that objects of similar size to them can be found further out from the Sun, in the region of the major planets. For the most part, these are satellites of the major planets. For example, Jupiter has four big satellites about the same size as the Moon and Mercury (together with a larger number of small satellites). The question, as with the Earth's Moon, is should these satellites be included as honorary members of the terrestrial planet group? Most astronomers would say no, mainly because an examination of the properties of these satellites shows that they differ from the terrestrial planets, especially in the materials of which they are composed. However, it is sometimes useful to bring in evidence from them when discussing the terrestrial planets. We shall find this necessary, for example, when we talk about the craters that have been observed on a number of planetary and satellite surfaces.

Let me go back briefly to the question of names. The four large moons of Jupiter are sometimes called the Galilean satellites. When the Italian scientist, Galileo, invented the astronomical telescope early in the seventeenth century, one of the first objects he observed was Jupiter. Though his telescope left a lot to be desired – it produced a much worse picture than a modern pair of binoculars – he managed to see the four main Jovian satellites. The point of this diversion is to illustrate another aspect of astronomical names. In order to understand them, it is often necessary to delve into history.

There is one planet we have not yet discussed – Pluto. This enigmatic body lurks near the far boundary of the major planets, yet is much smaller than the Earth. Like the Earth, it has a relatively large moon: so it might also be called a double planet. Despite its relative size, Pluto's moon was only discovered towards the end of the 1970s – a striking example of how difficult it is to observe this faint, distant planet. I remember presenting an early photograph showing the satellite to a meeting of the Royal Astronomical Society in London. Several of the astronomers present said they could not distinguish the moon from the general background blur of the photograph. It is correspondingly difficult to talk about Pluto's properties. However, it is clear that it has more in common with the large satellites of the major planets than with the terrestrial planets. A clue to Pluto's character may also lie in this planet's unusual orbit. With the exception of Pluto, all planets in the solar system pursue independent paths which never overlap each other. But Pluto has a clearly non-circular orbit which at its inner end overlaps Neptune's orbit. Indeed, though the textbooks usually refer to Pluto as the most distant planet, it recently cut across Neptune's orbit, leaving Neptune temporarily as the outermost planet. It has been suspected for many years that the orbit and properties of Pluto indicate that it is the escaped satellite of a major planet, rather than a truly independent planet. Though it is hard to establish whether this belief is correct, it seems reasonable to classify Pluto amongst major-planet satellites rather than with the terrestrial planets.

We have now had a brief look at all the planets and their satellites and it is time for another diversion on names. Planets are called after the gods of classical mythology. For the planets visible to the unaided eye – from Mercury to Saturn – the association of their names with gods and goddesses dates back to prehistory. The first planet discovered in historical times was Uranus, towards the end of the eighteenth century, and its naming raised problems. Its discoverer was William Herschel, a German resident in England, and he proposed to call the planet after the king of Great Britain, George III. But this was the time of the American War of Independence, and King George was unpopular abroad. After years of controversy, it was finally agreed to stick to tradition and use a name from classical mythology – in this case, Uranus. The same practice was followed for the subsequent discoveries of Neptune (in the nineteenth century) and Pluto (in the twentieth century).

Where possible, the names chosen for planets and satellites are meant to have some kind of relevance, however tenuous. For example, the name Uranus was chosen because the new planet lay just beyond the orbit of Saturn, and in classical mythology Uranus was the father of Saturn. Similarly, the two moons of Mars are labelled Phobos (closer to Mars) and Deimos (further away). This was thought appropriate because Phobos and Deimos were companions of the god Mars in classical tradition. The space age has created a problem by uncovering a range of new satellites of Jupiter and Saturn. Although the ancient gods were highly prolific in offspring, it is now becoming increasingly difficult to find appropriate names. Recently discovered satellites rejoice at least temporarily in such names as 1980 S26 (a satellite of Saturn discovered by one of the US Voyager space probes as it flew past the planet). The question of more romantic names is currently being pondered.

We have noted that, though some of the satellites of major planets approximate in size to the terrestrial planets, the two groups can be distinguished by their intrinsic properties. It is true, more generally, that all the groups of bodies in the solar system can be distinguished by their characteristics, as well as by their orbits. We have seen that the major planets are much bigger than the terrestrial planets. But the two groups also differ in the materials of which they are made. The Earth consists of a central core, mainly made of iron, which is surrounded by a thick mantle of rock. Jupiter, in contrast, is made up predominantly of the light gases, hydrogen and helium. The large satellites of the major planets (and Pluto) lie between these extremes. They contain rocky material, but this is usually combined with significant quantities of ice, which is considerably lighter. These differences in composition show up clearly if you compare the densities of the different planets. To find the density, all you need to do is to weigh a lump of the planetary material and compare it with the weight of an equal volume of water. The ratio of the two corresponds to the density of the material. Even though chunks of planet are not generally available on Earth, you can usually estimate the density from ground-based observations of the planet. The average density of the Earth is 5.5 (i.e. an average sample of the interior weighs 5.5 times the weight of the same volume of water). Ganymede, one of Jupiter's moons, has a density of about 2; whilst Jupiter itself has a still lower density of 1.3.

One important property of bodies in the solar system does depend on their distance from the Sun – the temperature reached by their surfaces. The Sun provides the heat to warm the exterior of these bodies, so we would expect bodies close to the Sun to have hot surfaces, whilst those at greater distances should be cooler. This is the way it works out in general (though, in detail, the way the Sun heats a surface can be more complicated). Mercury, the nearest planet to the Sun, has a surface temperature that can reach over 400°C (750°F) at its hottest point, while in contrast Pluto, the furthest away, is estimated to be at some 230°C (450°F) *below* the freezing point of water. But this is not the entire story. It has been found that at least three major planets, Jupiter, Saturn and Neptune, have temperatures that are higher than would be expected in terms of their distances from the Sun. The implication is that these planets are giving out appreciable amounts of heat from their interiors – something which no terrestrial planet does.

The ability to produce heat is a property usually associated with the Sun. To this extent, the major planets might be seen as scaled-down versions of the Sun. Moreover, a comparison of the Sun and Jupiter shows that the two bodies are made up of much the same materials – mainly hydrogen and helium. This faces us with one of the basic paradoxes of the solar system. The Sun at the centre of the system has certain properties – such as its composition and heat-emitting capability – which are mimicked in a pale way by the major planets. But between the Sun and the major planets come the terrestrial planets which have entirely different properties. Why are the planets separated out in this peculiar way? We shall see that this is one of the important questions to which the smaller bodies in the solar system can help provide an answer.

We have talked a good deal about how far away the planets are from the Sun. If on our trip into the solar system we had looked at how the planets are spaced relative to each other, it would have been immediately evident that the distance between planets increased rapidly the further we move away from the Sun. Some two hundred years ago a simple relationship was found between these planetary distances. Its discoverer was an obscure German called Titius, but it was publicized by his fellow countryman, Bode. The relationship is therefore sometimes called the Titius-Bode law, but it is more frequently known simply as Bode's law – an interesting example of the advertiser winning out over the inventor. The law worked well for Uranus, which was discovered shortly afterwards, but was subsequently found to break down for Neptune and Pluto.

In Bode's law, you start with the series of numbers 0, 3, 6, 12, 24, 48, 96, etc. (doubling the previous number each time except the first). Then you add 4 to each of these numbers, and finally divide by 10. The resulting series of numbers is: 0.4, 0.7, 1.0, 1.6, 2.8, 5.2, 10.0, etc. The interest of this result can be seen from the following table, which compares these numbers with the observed distance of the planets from the Sun (in Astronomical Units).

Planet	Mercury	Venus	Earth	Mars	?	Jupiter	Saturn
Distance (AU)	0.4	0.7	1	1.5	-	5.2	9.5
Bode's law	0.4	0.7	1.0	1.6	2.8	5.2	10.0

The law obviously fits the observations well, and shows very clearly how the gaps between neighbouring planets increase rapidly with distance from the Sun. The thing that puzzled astronomers in the eighteenth century was the significance of the apparently empty region at 2.8 AU. A major search was started for objects in this region and, on the first day of the nineteenth century, the largest of the asteroids, Ceres, was discovered at almost exactly the predicted distance. However, Ceres proved not to be alone: within a few years three more asteroids were discovered, and thousands have been seen since.

The name *asteroid* was coined by the same William Herschel who discovered Uranus. It is not a totally appropriate word, since it means 'star like' and, whatever else asteroids may be, they are obviously not stars. Herschel derived the name from the appearance of these objects as seen through a telescope. Unlike planets, which generally show a disc, the asteroids appear as points of light, like stars. The difference stems from the fact that asteroids are much smaller bodies than planets. This also explains, of course, why it took so long to detect them. Because the name asteroid does not seem entirely appropriate, some astronomers prefer to call these objects *minor planets*. The problem is that this sounds like the opposite of major planets, whereas, as we have seen, the true contrast is between major planets and terrestrial planets.

The first discovered asteroid, Ceres, received its name in the approved way for planets. Its discoverer was a Sicilian and Ceres was the local patron goddess in classical times. As the years passed it became evident that there were far more asteroids than could be given classical names. They are now usually called after more mundane creatures – human patrons, astronomers and even pet dogs. The naming system has also become more complex. On first sighting, an asteroid is now labelled by the year and a couple of letters that indicate when in that year it was discovered (e.g. 1983 TB was an asteroid discovered in October 1983). When it is clear that the asteroid's orbit is properly known, so that it can always be found again, it is assigned both a number and a name. In times past, many asteroids were found, but then lost again. It is the numbered asteroids (that is ones with well-determined orbits) which are usually considered to be the 'known' asteroids and some three thousand of these now exist. Often nowadays an asteroid is called by both its number and its name (e.g. 1 Ceres, 4 Vesta, 624 Hektor). The number is some indication of when the asteroid was discovered, and so of how bright it becomes.

Ceres dominates the asteroids. It is believed to possess nearly a third of all the material contained in asteroids but, even so, it is much smaller than our own Moon, containing only about 2 per cent of the amount of material in the Moon. In terms of size, Ceres is 600 miles (1000 km) across or more. Most of the asteroids that have been discovered since Ceres are not just smaller, but much smaller. There are hundreds of asteroids 10 to 20 miles (15 to 30 km) across.

Obviously, there would be problems if all these fragments circulated at exactly 2.8 AU. In fact, most (though by no means all) are spread out over a range of distances between 2.2 AU and 3.3 AU. This region is referred to as the *main belt* of asteroids.

It is not only in the size of the orbit – that is in the distance of the asteroid from the Sun – that differences can occur. Asteroid orbits can also be inclined at different angles to each other. If you imagine the Sun as a ball resting on the top of a table, the planets would be small marbles on the table top, all rolling round the Sun in the same direction. The important point is that they can be represented as more or less moving in the same plane – the table top. An astronomer looking for a planet knows that it will always be found in a narrow region of the sky, called the *ecliptic*, which corresponds to this plane seen edge on. Many asteroids could be similarly represented by tiny marbles rolling on the table top. But some would have to be shown looping up and down, above and below the table, and these are said to have orbits that are inclined to the ecliptic.

Some asteroids have orbits that differ in another way. We have talked about planets moving round the Sun in fairly circular paths (though we have seen that this is certainly not true of Pluto). Many asteroids also follow circular paths, but a significant number do not. In particular, about one in twenty has a noticeably elongated orbit, with the far end a good deal more distant from the Sun than the near end. Such asteroids can move out of the main belt and pass through the inner solar system in their wanderings. In doing so, they may cross the orbits of some, or all, of the terrestrial planets. A number cross the Earth's orbit, and so come closer to us than any planet can. We will be looking at these *Earth-crossers* in greater detail later, since they may have played a significant role in the history of the Earth.

As we have seen, our trip into the solar system would need to bring us in very close if we were to spot the asteroids. But there is another group of bodies that we might have come across much earlier – the comets. It is still not clear exactly where comets originate, but it is quite certain that they spend most of their time beyond the limits of the planetary system. They may even exist as far as half way out to the nearest star – a distance of some 100,000 AU from the Sun (remember that the most distant known planet is at only 40 AU). Even if this is an over-estimate, comets would obviously be the first solar-system objects that an incoming voyager could encounter. Whether they would be noticed is another matter. From Earth they are totally invisible unless, for some reason, they fall in towards the Sun and so pass through the planetary region.

A comet out in the depths of space is thought to be a small body, perhaps 10 miles (15 km) across or less. But this *nucleus* is not what we see when a comet dips in towards the Sun. As it approaches the Sun, the comet flares up into a large fuzzy ball – called the head, or *coma*. Material from this head streams back away from the Sun into a *tail* which increases in length the nearer the comet is to the Sun. A really long comet tail can be the largest thing visible in the solar system:

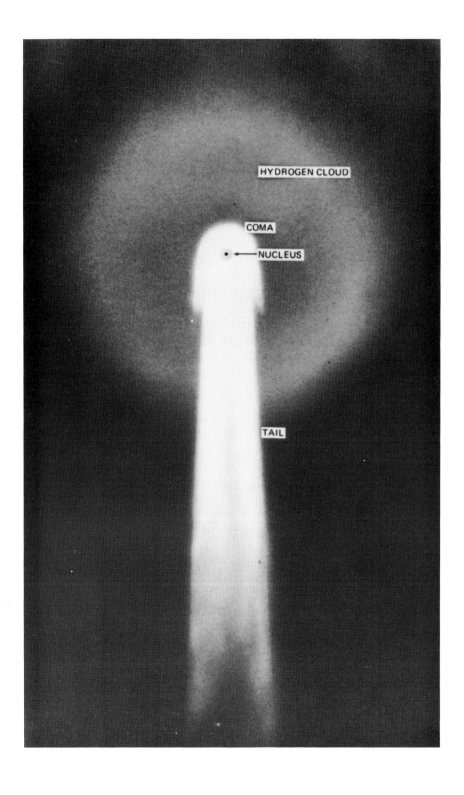

much longer than the distance from one side of the Sun to the other.

Comets fall in towards the Sun from all directions. Their orbits, unlike those of asteroids, show no sign of clustering round the ecliptic. Similarly, whereas asteroids move round the Sun in the same direction as all the planets (they are said to have *direct motion*), comets pass equally readily round the Sun in either direction. Many therefore have direct motion, but the remainder, going the other way round, are said to have *retrograde motion*. The deduction from these random motions is that comets probably form a cloud which surrounds the Sun in all directions. Unlike other solar-system objects, they are not concentrated in the plane of the ecliptic.

Because comets appear unexpectedly and anywhere in the sky, they are usually discovered by accident and the way in which they are recorded is a consequence of their haphazard nature. Immediately a comet is found it is given an appropriate letter. The first comet discovered in 1985 becomes 1985a; the second discovery is 1985b, and so on. To make the final result rather more systematic, the same comets are subsequently relabelled with Roman numbers according to their order of passage round the Sun. The first one to pass close to the Sun in 1985 therefore becomes 1985 I; the second is 1985 II, and so on. The numbers do not necessarily follow the same sequence as the letters. For example, comet 1985b might be discovered after it has passed the Sun, whilst 1985a is found before its close approach to the Sun. In that case 1985b becomes 1985 I, whilst 1985a is 1985 II. Fortunately, it is not usually necessary to bother about these complications. Comets are typically called after their discoverers and we can use these names, instead of the letters and numbers. For example, a bright comet was discovered by Morehouse in 1908. It was initially labelled 1908c, but subsequently relabelled 1908 IV. However, it is usually simply known as Comet Morehouse.

There is one qualification to add to our description of comets. A few of those which fall towards the Sun happen to pass close to a planet on their way in or out. The gravitational pull of the planet can change the comet's orbit until it follows a considerably different path round the Sun and, in some cases, the comet can be captured – that is retained within the planetary system – rather than return to outer

3 The parts of the comet an observer on Earth can see are the coma and (usually) a tail streaming away from the Sun. The hydrogen cloud can only be observed from satellites above the Earth's atmosphere. The word 'nucleus' is confusing. It is sometimes used to mean a bright sphere of gas and dust near the centre of the coma: at others, it means a small solid core, also at the centre of the coma, from which the cometary gas and dust originate. In the former sense, it is fairly commonly observed. In the latter sense, it has seldom or never been observed: the small dot inserted in the diagram shows the maximum relative size of a solid nucleus.

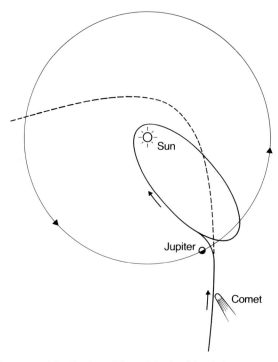

4 *A comet is captured by Jupiter. The original orbit of the comet round the Sun is indicated by the dotted line. The comet has happened to cross the orbit of Jupiter when that planet is passing close by and Jupiter's gravitational pull has drawn the comet into a new orbit round the Sun (indicated by the continuous line). The original path corresponds to a very long time for each orbit – perhaps millions of years – whereas the new orbit may take only a few years. The comet has changed from being long-period to short-period.*

space. Since the planets move round the ecliptic, such captured comets also usually move close to the ecliptic.

In terms of their orbits, therefore, there are two types of comets, usually referred to as *long-period* and *short-period* comets. The long-period comets are the typical objects described above which, because they have come from a great distance, may take millions of years to dip into the centre of the solar system and out again to the periphery. The captured short-period comets may require only a few years, or tens of years, to orbit the Sun once.

Since Jupiter is by far the most massive planet, and gravitational pull depends on mass, it is much more efficient at capturing comets than other planets are. So many captured comets start their circuits of the Sun from somewhere just within the orbit of Jupiter. These short-period comets typically follow elongated orbits, crossing the region of the terrestrial planets in their course. We have seen that some asteroids follow rather similar tracks, so this raises the question of whether we can distinguish between asteroids and comets if their orbits are similar.

At first sight, this seems a straightforward question to answer. The two types of body look different, so how could they be confused? To explain why life is not quite so simple, we must now make a major diversion to discuss how astronomers measure the properties of the objects they study.

It is only rarely that direct contact can be made with objects outside the Earth. Nearly all astronomical observation therefore depends on analyzing the radiation – light, heat and so on – that comes to us from bodies in space. The Sun (and other stars) are self-luminous: we see them by the light they produce. All other objects in the solar system depend on the Sun's illumination, so what we see of their properties depends on the way they react to sunlight. Suppose, for example, we look at the sunlight reflected to us from the surface of an asteroid. We know the colour of sunlight from direct observation of the Sun and we can compare this with the colour of sunlight reflected from the asteroid. We then try to interpret any differences in the two colours in terms of the properties of the asteroid surface, knowing that different sorts of rock reflect sunlight in differing ways.

Though this method is simple in principle, it is decidedly complicated in practice. For example, the Earth's atmosphere acts as a semi-transparent blanket over our heads which affects all light reaching us from outer space. This is why the daytime sky is blue. The atmospheric gases allow red light to pass fairly easily, but impede the progress of blue light. In fact, the blue light is bounced about in the atmosphere until it seems to come at us from all directions. Correspondingly, much of the red light from the Sun comes straight through, as is particularly evident when you see the Sun rising or setting. What this means is that the colour of the Sun, as you see it, is affected by the Earth's atmosphere (and varies according to how high above your horizon the Sun is). The same is true of any other astronomical object so, to get your colours right, you have to make corrections to allow for atmospheric effects.

Astronomers have various methods for measuring colour. One is to use a series of colour filters, passing blue light, yellow light, etc., and to measure the brightness of an object through each filter. A comparison of how bright the object appears to be through the different filters gives a measure of its colour. The Sun, for example, is normally brighter when measured through a yellow filter than it is through blue or red filters because the Sun's intrinsic colour is yellow.

This method – usually called photometry – provides a reasonable overall estimate of colour, but astronomers often need to compare colours in more detail than filters will allow. After all, there are many different shades of blue or red, as anyone who buys dress materials or paints will know. It is sometimes necessary therefore to examine the entire spectrum of sunlight, corresponding to the continuous band of colour that appears in a rainbow. The instrument used to do this is called a spectrograph. What it does, in effect, is to act like a very large number of colour filters, each of which simultaneously isolates a slightly different colour of the spectrum. The result is a very much more detailed specification of colour than can be managed with standard filter photometry.

You might wonder why, if a spectrograph provides so much information, photometry is used at all to determine colour. As with most things in astronomy, the answer is that there is a trade-off involved. The more detail you want to see, the longer time it takes to record it. If you are looking at a faint object, it may take an excessively long time to obtain a spectrum with a spectrograph, whilst its overall colour can be measured quite rapidly using filter photometry.

When the sunlight reflected by asteroids is observed, it is found that the asteroid surface produces small changes in colour, which are spread out over fairly wide regions of the spectrum. This is typical of what you would expect for a solid surface. Comets produce a quite different result. A cometary spectrum shows bright light emitted at very precisely defined colours. These *emission lines*, as they are called, are characteristic of the light which gases produce when they are illuminated by the Sun: establishing the precise colour of the lines will tell us what gases are present. This observation helps us to understand why comets produce heads and tails as they approach the Sun. The gases present come from materials that would be in the form of ice in the outer solar system, including not only water ice, but also carbon dioxide ice (the stuff ice-cream manufacturers use to keep their product cold) and others. As such icy material nears the Sun, it evaporates to produce a cloud of gas, which expands to form the head and tail. This explanation means we must suppose that the original cometary nucleus contained a lot of ice.

After this long diversion, we can now return to the question with which we started – how can we distinguish between an asteroid and a comet in the same orbit? The simplest answer is that the spectrograph will tell us that an asteroid is a lump of rock whilst a comet is a lump of ice.

But using a spectrograph also reveals another characteristic of comets. If you look closely at the spectrum of a comet, you find that, beneath the bright emission lines, there is the same kind of reflected sunlight spectrum that asteroids produce. This means that comets also contain rocky material (though it seems to be mainly in the form of many small dust particles, rather than one large lump of rock). A comet following the same sort of orbit as an asteroid will be so close to the Sun that its ice will evaporate continuously. What will be left when all the ice has gone? It may be that there will be nothing left, but some astronomers believe from the evidence of the dust that the ice will leave behind a rocky residue. If so, they say, how could this core be distinguished from a small asteroid? This is a question we will come back to in the last chapter.

The presence of cometary dust raises another query. Cometary nuclei release dust in all directions, and this dust is hit by sunlight which exerts a very small pressure on it. Although the resultant push is so small that it requires quite sophisticated experiments on Earth to detect it, for particles in space the pressure can build up to produce a significant effect. Sunlight initially forces the dust backwards away from the Sun and into the tail of the comet, but the pressure continues to act, even in the tail, so the final result is that the dust is dispersed throughout the region traversed by the comet's orbit through the inner solar system.

The existence of dust in the space between the planets has been known for many years. Under favourable conditions, after sunset or before sunrise, it is possible to see a faint haze stretching along the ecliptic which is known as the *Zodiacal light*. This rather odd name stems from the fact that the ancient constellations of the Zodiac – Aries, Taurus and so on – lie along the ecliptic. Zodiacal light is simply sunlight reflected from the dust particles lying between the planets; but the question we have to ask is – does all this dust originate from comets?

Once again, though in a different form, we are faced with the problem of confusion with asteroids. The smallest asteroid we can hope to see from Earth in the main belt would be a few miles across. But there is evidence to suggest that material in the region of the asteroids comes in all sizes down to small dust particles. Might not interplanetary dust come from the asteroid belt as well as from comets? This is another query we must return to later.

Both the heads and tails of comets consist of a mix of gas and dust, so the gas is swept backward from the head in just the same way as the dust. Again, the Sun is the culprit, but the mechanism in this case is different because the pressure comes from a wind blowing through the solar system. To see why such a wind exists, we need to look again at the Sun. Like all stars, the Sun is hot – several thousand degrees Centigrade at the surface. Above this surface, a thin solar atmosphere stretches out into space which is normally too faint to be seen against the glare of the Sun; but it can be observed easily during a total solar eclipse, when the Moon blocks out the light from the Sun's surface. From observations under these circumstances, it can be deduced that the solar atmosphere is much hotter than the surface and is therefore expanding outwards away from the Sun. In doing so, it streams out past the planets and produces a wind – the *solar wind*. This wind sweeps away any gas present between the planets and, in particular, any gas evaporated from a cometary nucleus is swept backwards to form a tail (so a comet's tail always points away from the Sun).

Despite the sweeping action of the solar wind, there are always a few particles in the interplanetary gas which do not come from the Sun. The most interesting are the *cosmic rays*, which are electrically-charged particles moving at high speeds. In principle, solar-wind particles can be called cosmic rays, but here this term will apply particularly to particles that come from sources outside the solar system, although what these sources are is still a little doubtful. As we shall see in Chapter 2, electrically-charged particles can have their paths altered if they come near magnetic bodies. As both the Sun and the Earth are magnetic bodies, the cosmic rays are deviated so much that it is extremely difficult to pin down the original direction from which they came.

5 This picture of the Zodiacal light was taken high up in the Bolivian mountains. The Sun is below the mountainous horizon at the bottom and its light is being reflected to us from the dust that lies about between the planets. Because the amount of light is very small, the photograph had to be exposed for some time. In consequence, the background stars moved, owing to the Earth's rotation, and appear here as trails.

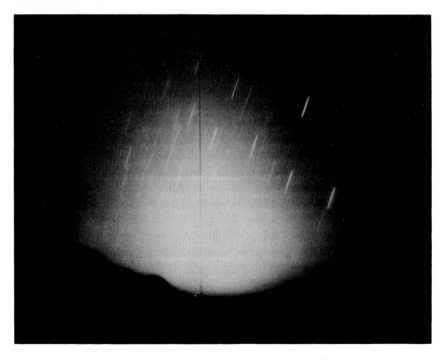

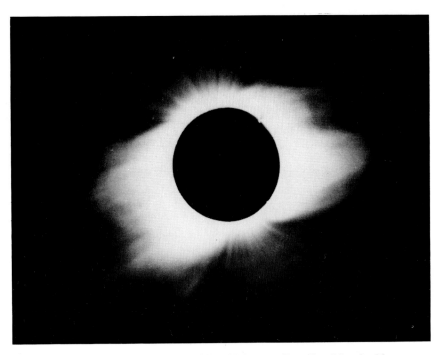

6 The Sun has been caught in the middle of being totally eclipsed by the Moon. Now that the glare from the Sun's surface has been removed, we can see its atmosphere stretching out into space. The north and south poles of the Sun are towards the top and bottom of the photograph. Each is marked by a fan of rays, caused by the Sun's magnetism. The Sun's equator slopes up from bottom left towards top right – in these directions, the solar material we can see is streaming out into the solar system.

The solar wind and cosmic rays both hit the Earth (with results that will be described in the next chapter), and so may some of the other bodies we have described, in particular asteroids and comets which cross the Earth's orbit. Fortunately, impact with the latter is unlikely: the odds on a rifleman hitting a pinhead at maximum range are much better. However, there are many more small fragments of rock flying round the solar system than there are larger bodies and the probability of one of these smaller pieces hitting the Earth is, therefore, much higher. A few thousand fragments, inches to feet in diameter, enter the Earth's atmosphere each year, whilst smaller particles – a fraction of an inch across – are entering all the time. These fragments are travelling at several miles per second relative to the Earth. When they plunge into the atmosphere the friction heats them up and they are seen as flashes of light across the sky. It is customary to distinguish between the larger incoming fragments, which reach the Earth's surface, and the smaller particles which burn up completely in the atmosphere. The former are called *meteorites*; the latter – *meteors*. If both groups are referred to simultaneously, they are called *meteoroids*.

These names seem to relate the objects to meteorology, rather than to astronomy, and this is because, centuries ago, the main interest in the atmosphere was in sudden changes. In particular, people looked for any flash of light – lightning strokes, for example – and called such events meteors. The study of the atmosphere therefore came to be called meteorology, but, oddly, the word meteor became narrowed down in its meaning till it only applied to *shooting stars*. Nowadays, a number of words may be used to describe the bright flash of a meteor across the sky: one that is particularly significant is the word *fireball*. In the next chapter we look at the possible implications of seeing a fireball.

2
An Earth-Based View

Just after New Year, 1966, rumours began in Leicester of strange lights seen in the sky over a wide area of the English Midlands on Christmas Eve. At the same time, a report came through that a number of rocks had fallen, apparently from nowhere, on a village south of Leicester called Barwell. My colleagues and I put these two pieces of information together and immediately organized an expedition to Barwell. We hoped the rumours meant that one of those rarely observed events – the fall of a meteorite – had occurred. (If that sounds a bit cautious, remember that strange sights and sounds are not uncommon on Christmas Eve.) Over the next few weeks we pieced together the following story.

Christmas Eve had been fairly cloudy over the Midlands that year, but the cloud was patchy: some regions had total cloud cover, whilst others had a nearly clear sky. In those areas where it was fine, many of the people out-of-doors had seen a bright fireball crossing the sky just after 4 pm. The sight was so unexpected, and passed so quickly, that few could remember in detail what they had seen. Yet we finally found out a surprising amount. For example, one man had been crossing a golf course and remembered that the streak of light touched the top of the flagstick on one of the greens. From that we managed to work out how high the fireball was above his horizon.

Our first impression when we put the reports together was one of total chaos. It soon became apparent that this was because there had been more than one fireball – we had been trying to combine all the observations as though they referred to a single fireball. We finally managed to reconstruct the tracks and these demonstrated that the first observations were of two fireballs on more or less parallel paths, moving in a north-northeast direction and coming down at an angle of about twenty degrees to the horizontal. The more westerly of the fireballs split into two south of the town of Warwick. The light from all three fireballs disappeared when they were between 5 and 6 miles (9 and 10 km) above the Earth's surface (that is, at round about the cruising height of a jet aeroplane). Before the fireballs disappeared, some observers noted that they were leaving white trails behind them, which persisted for a while.

The passage of the fireballs also produced sound. Noises were heard over a wide area – in a band 30 miles (50 km) wide along the final 90 miles (150 km) of the fireball tracks. Some observers reported what they heard as thunder, others as sonic bangs, while a few claimed that the sounds produced physical effects: for example, an electric light bulb breaking as the bang arrived.

Of the final three fireball tracks, only the one directed towards Barwell produced a fall of rocks. Perhaps the other two also deposited material, but it went unnoticed because the rocks fell in open countryside. People in Barwell who were out and about on Christmas Eve heard a swishing noise followed by a succession of dull thuds. Their reactions to what they had experienced varied. One man we interviewed had had the bonnet of his parked car hit by a falling fragment and he had initially been sure that it had been thrown by a couple of boys up the road. It did not cheer him to find he was wrong: vandalism came under his insurance cover, but a meteorite was reckoned an act of God. In general, surprisingly little damage had been done and this was largely because the main fragments had fallen to one side of the village centre, on playing fields and common land. One small piece had penetrated the roof of a factory building and the ceiling below and from the relative positions of the holes we could work out that the fragment had been coming down at an angle of about eighty degrees to the horizontal when it hit the building.

The various pieces of meteorite, when put together, weighed over a 100 lb (45 kg) – the most massive meteorite ever observed in the UK. We found a fair amount of this material and the remaining fragments were unearthed by local inhabitants and visitors in response to the offer of a reward. (One couple paid for part of their honeymoon with what they found.) The fragments were scattered over a roughly circular patch, with the smaller pieces falling on the southern side of this area. Although the original fireballs were travelling at high speed, the fragments must have been moving fairly slowly at the end. The ground was thawing after snow, yet the deepest any piece penetrated was only a foot or so and some fragments were found lying on the surface with the snow hardly disturbed. The fragments clearly came from a single body which broke up at a late stage in its descent: it was possible to fit them back together like a three-dimensional jigsaw puzzle.

7 RIGHT ABOVE *This small fragment of the Barwell meteorite has a black fusion crust on its top. If you look closely at the greyish interior, you can see fragments of some of the larger chondrules (see p. 76).*

8 RIGHT BELOW *A close-up of a Barwell fragment showing the change in appearance of the fusion crust on different sides of the meteorite. Because the meteorite maintained the same orientation during its flight through the atmosphere, one half of its surface experienced much greater heating than the other.*

I have described the events at Barwell in detail because they are typical of what you see when a meteorite hits the Earth. We can now go back and look at what was actually happening at each stage.

When a meteorite meets the Earth, it is usually travelling at a high speed, partly because all bodies in the inner solar system are moving rapidly. The Earth, for example, is moving round the Sun at about 20 miles (30 km) per second. But a meteorite is also helped by the Earth itself, which, through its gravitational pull, accelerates bodies towards its surface. These two reasons together mean that a meteorite usually enters the top of the Earth's atmosphere at a speed of several miles per second. At this speed there is a great deal of friction between the surface of the meteorite and the atmosphere. As the meteorite falls, the atmosphere becomes denser, and the friction is correspondingly higher. By about 100 miles (150 km) above the Earth's surface, the heat produced by the friction has become so intense that its effects cause the meteorite to emit light. Some light results from the now molten material on the meteorite's surface, but most comes from the heated atmosphere along the meteorite's track. The reason the atmosphere gives out more light than the surface of the meteorite is because more of it is affected. For example, if the meteorite is about the size of a football across, then the atmosphere it heats covers an area about the size of a football pitch. The light emitted as a result is seen from the ground as a fireball.

Although the molten surface of the meteorite may not be important in terms of light, it is significant in another way. The rush of air past the meteorite makes the molten material on its surface flow backwards, until it finally falls off the end as small droplets. These cool and solidify, rapidly forming a white tail (as we saw behind the Barwell meteorite). The tail disappears as the droplets spread out in the atmosphere and begin to fall to Earth.

The result of this process – which is called *ablation* – is that the meteorite rapidly becomes smaller. Thus a large meteorite which fell in Siberia in 1947 is estimated to have left behind a tail containing three times as much material as was subsequently picked up in the form of fragments on the ground. A small body can be completely melted away before it reaches the Earth's surface – if the meteorite is less than a few pounds in mass when it enters the atmosphere, it is unlikely that anything will subsequently be found at ground level. This question of whether a body burns away in the atmosphere forms the basis of the division mentioned in Chapter 1 between meteorites – bodies from outer space which deposit some remains on the Earth's surface – and meteors – which melt away completely in the atmosphere.

The word fireball is reserved for the brightest flashes of light across the sky (though some people prefer to call these *bolides* and to restrict fireballs to objects that are a bit less bright). This does not mean that all fireballs lead to the fall of a meteorite and, in fact, only a small proportion do. An important factor in a meteorite's survival is whether the body fragments. The air pressure not only produces heat, but also puts considerable stress on the incoming body, something that is clear from the sounds associated with fireballs (sonic bangs are a good indication that the air in front of a projectile is highly compressed). This stress usually leads to fragmentation, unless the body is very strong, so most meteorites come to Earth as an assortment of fragments, rather than as a single body. However, several small pieces ploughing through the atmosphere will melt away more rapidly than one large piece, so fragmentation is likely to reduce the chances that material will reach the ground.

As we saw, two fireballs were initially connected with the Barwell meteorite – suggesting that a single body broke up high in the Earth's atmosphere. When one of these fireballs later split into two, this showed that there had been another break up at a lower level. A final split after the fireballs were extinguished can also be deduced from the appearance of the fragments picked up at Barwell. The light from the Barwell fireball disappeared at a height of some 5 miles (8 km) above the Earth's surface, indicating the point at which atmospheric friction had slowed down the incoming body so much that little new heat was being generated. Consequently, the fireball went out, and at the same time the molten layer on the surface of the meteorite solidified into a thin black crust over the whole of the rock. But the fragments picked up at Barwell were not completely covered by such a crust. Instead, a number of them had a crust on one side, with the other surfaces obviously unheated. This shows that the final fragmentation occurred after the fireball was extinguished.

The molten surface of the Barwell meteorite solidified quickly. This has helped preserve evidence which confirms our picture of what happened to the rock during its flight through the atmosphere. The black crust has channels in it where the molten fluid flowed away from the front end of the meteorite. It is even possible to see some of the small droplets which were about to be blown off the meteorite and the results of air friction on the same front face. If a meteorite keeps the same face forwards during its flight, this face is eaten away by air friction until it assumes a cone shape and this is what happened to the Barwell meteorite in the last stages of its path through the atmosphere. (If the meteorite is spinning, material is removed from all sides, so that no particular shape results.)

Almost all the speed the Barwell meteorite had in space was destroyed in the atmosphere, so that the final fragments fell nearly vertically and quite slowly; as we were able to deduce from the positions of the holes in the factory roof. However, a little forward speed remained, and, in consequence, the heavier fragments were carried further along the track of the fireball (because the air resistance had less effect on their motion). This is why the smaller fragments were found together on the south side of the Barwell fall.

At this point, I shall deviate briefly to talk about the naming of meteorites. The simple rule is that the meteorite is called after the nearest named geographical feature. So I have been talking about 'the Barwell meteorite', and that is how it is labelled in museums. The method is straightforward, but there can be complications. Many meteorites break up to an even greater extent than the Barwell meteorite, and the pieces may spread out over an area containing several different geographical features. Which feature should the meteorite be called after? In the past, this difficulty has led to several different names being applied to the same meteorite – a source of endless confusion – but now the names have to be agreed internationally. The British Museum (Natural History) in London has long been involved in looking after this activity and I remember visiting the man in charge some years ago. He was shaking his head and saying, 'No, we can't call it this.' The problem was that a meteorite had recently fallen in the USA near an area called the Bloody Basin. However, a glance at his catalogue shows that he gave way in the end – there is the Bloody Basin meteorite.

The sights and sounds at Barwell were typical, but many meteorites fall with no one to observe them, for much of the Earth is ocean, or sparsely populated. Even if the fireball is widely seen, it is usually a matter of luck whether any meteorite fragments are found. Only half a dozen meteorites are picked up in the average year – perhaps only one in a hundred of those that fall. Then again, as we have said, not all fireballs produce meteorites. Whether they do depends mainly on the material of which the meteorite is made, particularly its resistance to fragmentation. We shall look at the differences between meteorites again in Chapter 4, but in terms of strength three groups can be distinguished. The toughest group is made mainly of iron, though some nickel is also present. Sometimes these *irons* are mixed with rocky material and they are then called *stony irons*. Another group covers various mixtures of rock: its members are said to be *stones*. The final group covers the rather fragile *carbonaceous meteorites*. These often contain a fair amount of rock, but they also have an appreciable proportion of material made out of carbon. This tends to be black, a bit like compressed soot.

I once saw a small boy trying to carve his initials into a large iron meteorite on display in the British Museum (Natural History). Predictably, the iron was so hard that the blade of his penknife broke. If the meteorite had been a stone he might have managed to scratch it with some graffiti, while, at the other end of the scale, some

9 ABOVE *The fireball track in the photograph overleaf led to the discovery of the meteorite that caused it, near Lost City in Oklahoma. The Lost City meteorite is essentially a small piece of rock (a stony meteorite). Its size indicates how little material is needed to produce a great amount of light during descent through the atmosphere.*

carbonaceous meteorites are so soft that a penknife could carve them as easily as butter. The result of these differences in strength is that irons are the most likely to survive a passage through the Earth's atmosphere, whereas the carbonaceous meteorites are most likely to fragment and perish.

These differences in strength also result in a misleading impression of the frequency with which different kinds of meteorite hit the Earth. Observed falls of meteorites are rare: I regard myself as very fortunate to have been involved in two – one in the USA and the Barwell fall in England. But, clearly, it is possible to find a meteorite fragment on the ground long after the fall has occurred. Meteorites which are discovered in this way, without being seen to fall, are called *finds*. H. H. Nininger, the most persistent of meteorite hunters, quotes the following example:

10 OVERLEAF *A fireball passing over the USA: some idea of its brightness is given by comparing it with the streetlights in the bottom corner. The stars appear as trails because the camera had been left open for some time in the hope of detecting any fireball tracks as they occurred. This was one of the rare times when that hope was fulfilled. The breaks in the fireball tracks are not real: they are produced by a rotating shutter in the camera. Their presence makes it possible to determine how fast the fireball is moving.*

When the first cattlemen settled in what is now Kiowa County, Kansas, their attention was drawn to a few heavy black rocks scattered here and there over a small area near the present location of the little town of Haviland. These 'rocks' attracted attention, not only because of their great weight and peculiar colour, but for the additional reason that no other stones of any kind could be seen for miles around.

The evident peculiarity of these 'rocks' was due to the fact that they were actually iron meteorites. Stony and carbonaceous meteorites are less obviously different. Moreover, water and frost erode away stony and, especially, carbonaceous material much more rapidly than they affect a mixture of iron and nickel. Consequently, the vast majority of finds are irons. Museum collections around the world, which contain both falls and finds, give the impression that iron meteorites are the commonest, followed by stones, with carbonaceous meteorites a long way behind. When allowance is made for atmospheric effects, and for the relative likelihood of the different meteorites being found on Earth, it is reasonable to suppose that the order is the other way round – with carbonaceous meteorites the most common, and irons the least.

In recent years, Antarctica has become a favourite hunting ground for meteorites. Out on the ice sheet, a meteorite of any type looks different from its surroundings, while the low-temperature conditions also mean that erosion tends to be slower. Moreover, the Antarctic continent itself helps hunters, because the slow motions of the glaciers concentrate the meteorites into smaller, more easily explored regions. Glaciers are like slowly flowing rivers. Just as rivers throw up debris at particular points along their banks, so glaciers preferentially deposit meteorites at specific points along their courses. Several hundred meteorite fragments have been retrieved from Antarctica in the last few years. These range over a wide variety of meteorite types, including some entirely new ones: they have shown how incomplete our knowledge of meteorites still is.

All this means that meteorites are eroded in two ways – by the atmosphere, and by forces of erosion on the Earth. We can get some idea of the amount of erosion for a given meteorite by using an effect produced by cosmic rays.

As we saw in the first chapter, cosmic rays are electrically-charged particles moving very rapidly. When these high-speed particles hit a meteorite in space, they can react with the atoms it contains, turning some into different atoms and making others radioactive. (A radioactive atom is one that breaks down sooner or later into a different type of atom.) Some of the new types of atom are unusual – that is, not commonly found in nature – and so can be detected fairly easily. In space, the interior of a meteorite is partly protected from cosmic rays by its own outer layers and, consequently, the effects of

11 *So far as I know, the character in this strip – BC – has never been into space. If he had, he might find that astronauts, no longer shielded from cosmic rays by the Earth's atmosphere, would agree with him.*

the cosmic-ray bombardment fall off from the surface of a meteorite towards its centre. Laboratory measurements of unusual atoms can therefore be used to draw a kind of contour map showing how far any point was below the original surface of the meteorite when it was in space. We can work out how material was melted away during the descent by looking for gaps in these contours.

The radioactivity produced in meteorites also has its uses. Once a meteorite has landed on the Earth's surface, it is shielded by the atmosphere from any further bombardment by cosmic rays. As a result, the amount of radioactivity dies away gradually with time, so what is left can be used as a measure of how long the meteorite has been on the Earth's surface. Some of the iron meteorites that have been found have been lying around for up to a million years, but carbonaceous and stony meteorites disappear in a much shorter period. These cosmic-ray measurements give a good idea of how well different meteorites resist terrestrial erosion.

There is some further information that can be got from looking at cosmic-ray effects. The amount of cosmic rays that hit a body depends on its distance from the Sun, so a meteorite's radioactivity can tell us something about its path through space before it hit the Earth. This method is hard to apply in practice, but it has been used to show that meteorites typically follow elongated orbits round the Sun. At one end of the orbit, the meteorite moves inside the Earth's orbit and gets closer to the Sun. At the other end, it reaches the region of the asteroid belt. This picture ties in well with what is known about the orbits of meteorites from direct observation.

It might seem straightforward to observe the path of a fireball and then calculate from which direction in space it appeared. The problem is that the observed path is affected by both the resistance of the atmosphere and the gravitational pull of the Earth. If you are going to allow for these effects, you need very accurate information on the meteorite's path through the atmosphere. To date the most satisfactory way of plotting this has been to take photographs of the fireball with special cameras at a number of widely separated sites. The obvious problem is that meteorites arrive unexpectedly, so photographing them in this way poses particular difficulties.

Two detailed studies of fireballs have been carried out since World War II and in each case a network of cameras was installed and photographs were taken automatically in the hope of, sooner or later, catching a fireball. One of these networks was set up in the American Mid-West; the other was in Eastern Europe. As we have seen, the number of bright fireballs is small, and the number that lead to the recovery of meteorites is even smaller. After several years of effort, orbits were determined for just three meteorite falls, all of which, as Figure 12 shows, originated in the region of the asteroid belt. In fact,

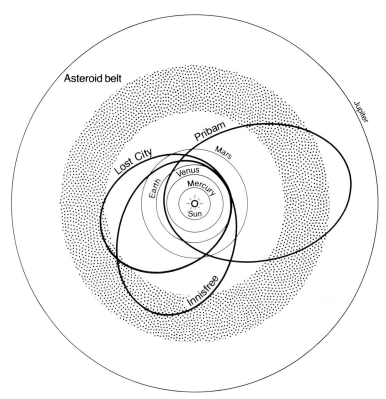

12 To obtain accurate tracks of meteorites before they hit the Earth's surface requires precise photographic measurements. As a result very few reliable orbits are known: the three best are shown in this diagram, all of which originate in the region of the asteroid belt. Innisfree refers to Innisfree, Alberta; Lost City to Lost City, Oklahoma; Pribam to Pribam, Czechoslovakia.

their paths are very like those of Earth-crossing asteroids, such as the orbit of the asteroid Apollo.

There is one point we should look at again here. As we have seen, much of the material of a meteorite is blown off as molten droplets during its passage through the atmosphere and, because they are small, these droplets cool rapidly and also slow down quickly. After a short time, they become solid, glassy spheres drifting down under the pull of gravity towards the Earth's surface. These small spheres are widely distributed across the Earth: they are found, for example, buried in the sediments at the bottom of the oceans. One of my colleagues has even found them embedded in lumps of salt from salt mines where they must have been trapped long ago as the salt crystallized from water. A fireball that fails to deposit a meteorite may well leave deposits of this sort.

We must now look again at the apparently arbitrary distinction we drew in Chapter 1 between meteorites and meteors. After all, on a

body like the Moon, which has no atmosphere, all incoming matter hits the surface and so is *meteoritic*. But this is not the whole story. The Earth's atmosphere acts as a kind of filter: material that melts more readily, or is fragile, breaks up in the atmosphere and is removed, while stronger and less-easily-melted material can survive. Some types of material may simply be incapable of withstanding atmospheric effects, and so never be able to form a meteorite. If, for example, a snowball hits the top of the atmosphere, it is long odds against it ever reaching the surface. This is not a far-fetched example, for if you look at the way meteors burn up in the atmosphere, it appears that some actually do have some properties rather similar to snowballs. (The main difference is that they are much smaller than anything we would call a snowball: meteors tend to be only a fraction of an inch across.)

Oddly enough, it is easier to work out where some of these meteors come from than it is to determine the orbits of meteorites, even though the meteors never reach the surface. The reason is that some meteors appear in groups and every so often a whole group will hit the Earth's atmosphere in a short space of time. Such an event is called a *meteor shower*. With a really numerous group, the appearance of the sky may actually give the impression of a fiery shower. A contemporary description of the great meteor shower of 1866 said: 'People of Beirut saw thousands of these meteors, mixed in commotion and confusion, and they compared their extent in the heavens to the spreading out of locusts in the sky.' Less striking showers can be seen every year, most of them repeating at about the same time of the year. For example, a small shower is visible every December and the meteors it produces are called the Geminids.

The naming of meteor showers is simple. If the paths of the individual meteors making up the shower are traced, they are found to radiate out from a specific region of the sky and the shower is called after the constellation where this *radiant* occurs. For example, the Geminids seem to shoot out from the constellation Gemini (the heavenly twins).

The impression that the meteors are spreading out from a particular part of the sky is actually due to perspective. The effect is similar to looking at two parallel lines on Earth – two railway lines, for example – which appear to meet at the horizon. In the case of meteors, there is a stream of particles moving round the Sun in parallel paths, and as they come into the Earth's atmosphere, their trails allow us to look back along their incoming tracks. The middle of these paths – the radiant – fixes precisely the position of their orbits in

13 This drawing of a fireball over Madrid in 1896 shows how the build up of atmospheric pressure can often cause an incoming body to fragment.

14 The tracks seen here represent the burn up of dust particles from the Leonid meteor shower as they hit the Earth's atmosphere. The Earth intersects this stream every November. The meteors appear to be coming from a point above the top right-hand corner of the photograph; this point lies in the constellation of Leo, from whence the shower gets its name.

space. On the basis of the size of their orbits, meteor showers may come from further out in the solar system than meteorites. However, this is not always true: the Geminids only go out as far as the asteroid belt.

The rather limited information we have on meteorite orbits seems to link meteorites with the asteroid belt, while in contrast the information we have on orbits of meteor showers links them more frequently with comets. For example, the Leonids observed in November 1866 were found to be following the same track as a comet discovered in January of the same year. It follows that the line we have drawn between meteors and meteorites may actually be of considerable significance, as it may also distinguish between asteroid-related and comet-related objects, a conclusion that we shall return to again later.

We cannot handle most of the solid material hitting the Earth because the atmosphere prevents it from reaching the surface, so one way of getting round this problem would be to examine the material whilst it is outside the atmosphere – using either manned, or

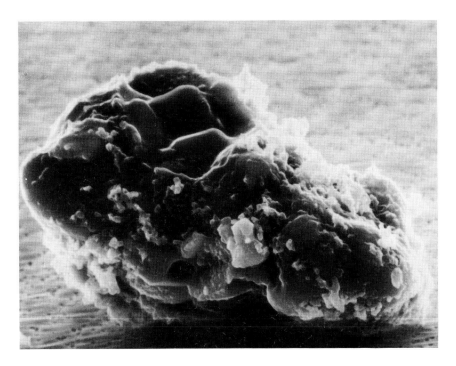

15 A dust particle from interplanetary space shown magnified by a factor of about 15,000. It was collected as it drifted down through the Earth's atmosphere by a high-flying aircraft belonging to NASA (the National Aeronautics and Space Administration).

unmanned satellites. Unfortunately, most such attempts are self-defeating. The satellites are travelling with the Earth, so the material they are intended to examine will be moving at high speed relative to them. Any attempt to collect pieces would involve an impact that would probably destroy the material. However, if the incoming particle is small enough, the Earth itself can again act as a collecting device. We saw that small spherules blown off from meteorites float down quite slowly to the ground, because they are much more affected by air resistance than larger fragments. (If particles are small enough they can take weeks, or even months, to settle to the surface from the upper atmosphere.) Similarly, very small dust particles from interplanetary space are slowed down rapidly when they hit the atmosphere and because they are so small this will happen before they become too greatly heated and melt. Consequently, a shower of interplanetary dust – made up of particles only a small fraction of an inch across – is continually falling to the Earth's surface. It would be difficult to detect these particles in the lower atmosphere, because this contains considerable quantities of ordinary terrestrial dust scooped up from the surface, but high in the atmosphere extra-terrestrial dust should be commoner than terrestrial dust. A high-flying jet should therefore be able to collect interplanetary particles and bring them back for study in the laboratory. This has now been done many times by researchers in the USA and it has been found that the extra-terrestrial particles are quite easy to distinguish: at this height, most of the other solid particles present come from rocket exhausts.

The most important result of this work has been to show that the dust entering the atmosphere has two basic types of structure. Some is in the form of solid grains, very like tiny meteorites, but some has a highly porous structure with individual particles often looking like a bunch of grapes. If larger bodies with this type of structure hit the atmosphere, they would fragment and burn up rapidly and we would be most unlikely to recover pieces on the surface. One possibility is that the solid grains came originally from asteroids, whereas the porous particles came from comets.

We have restricted ourselves so far to talking about solid bodies that hit the Earth, but in Chapter 1 we saw that there is also gas in the solar system, blowing outwards from the Sun. It is an interesting question whether any of this gas is picked up by the Earth. Unfortunately, though the question seems simple, the answer proves to be complicated because of the problems posed by the nature of the gas and the peculiarities of the Earth.

The gas from the Sun is very hot. At high temperatures most of the electrically-neutral atoms that would normally make up the gas are broken down into particles with either positive or negative electrical charges. A gas which contains many such charged particles is called a *plasma*: so we can say that the solar wind represents an outflow of plasma from the Sun. (The word plasma, like meteor, has to be treated with caution. It is used in a quite different sense in biology and medicine.) Since a plasma contains equal numbers of positive and negative particles, it is electrically neutral overall, so what difference do the charged particles make? We can see the answer, by analogy, in the corresponding properties of solids. There are two major types of solid – metals (e.g. a copper wire) and insulators (e.g. a piece of wood). Both are electrically neutral, yet their properties are obviously different: metals can conduct electricity and can be affected by magnets, whereas insulators cannot. We can equate a neutral gas to an insulator and a plasma to a metal in terms of their reaction to electricity and magnetism. This difference becomes important when we look at the way the solar wind interacts with the Earth.

The Earth, unlike other planets nearby – Mercury, Venus and Mars – is highly magnetic. The impact of the solar wind on these other planets is straightforward because the particles hit the atmospheres of Venus and Mars, and the surface of Mercury (which has no atmosphere) directly. What happens with the Earth is quite different, and much less easy to understand. One of the examiners for my doctorate about a quarter of a century ago was Professor Vincenzo Ferraro; he had been one of the pioneers in the early 1930s who had studied the interaction of the solar plasma with Earth. After my examination he told me that still, thirty years on from his first studies, he was far from clear about all the processes involved. Subsequent direct investigations by satellite have shown how right he was to hesitate over the complexity of solar-terrestrial relationships. But his work, and that of many others, has at least made the main outlines reasonably clear.

We normally think of the Earth's magnetism as the force which makes a compass needle point towards the north. (In fact, as we are always reminded on maps, the magnetic north pole does not coincide with the geographical north pole, so the compass needle points a little to one side of true north.) What we tend to forget is that the Earth's magnetism also operates vertically. At the magnetic north pole, a compass needle free to move in any direction would point vertically downwards; at the magnetic equator it would lie parallel to the Earth's surface. It is now possible to map the Earth's magnetism not merely on the Earth's surface, but also out into space by putting appropriate measuring instruments into satellites. The results are shown in Figure 22 (see p. 64). The lines show the direction in which a compass would point, if it was placed at the corresponding position in space. Although these lines are obviously fictional, it is nevertheless convenient to use them in describing how the solar plasma is influenced by the Earth's magnetism simply because, whereas neutral atoms are not affected by magnetism, electrically-charged particles tend to move along these imaginary lines rather than across them. (In other words, charged particles prefer to move backwards or forwards along the direction that a compass needle points.) The Earth's magnetic effect extends into space far beyond the terrestrial atmosphere, so the plasma from the Sun encounters this magnetism first. The charged particles are affected by it in such a way that they are deflected round the Earth (see Fig. 22), the picture that results being rather like the flow of water round a stone in a stream. Also, like flowing water, the solar wind exerts a pressure on any obstacle to its progress, so it compresses the Earth's magnetism on the side towards the Sun. Correspondingly, there is a lower than usual pressure on the

16 OVERLEAF *One of the typical forms of aurorae that are common near the north (and south) magnetic pole. The 'curtains' look as though they are quite close, but are actually high in the Earth's atmosphere.*

downwind side, and this produces an elongated tail. The solar wind therefore determines how far into space the Earth's magnetism can stretch: the region it defines around the Earth is called the *magnetosphere*. (It is evident from the diagram that the region is far from spherical. The word is being used here in the same sense as 'sphere of influence'.)

We can now return to our initial question – how much gas is captured by the Earth? From what has been said so far, it might be supposed that the answer is none because the solar wind flows past the Earth. But this simple picture is inadequate. If the plasma flows along the lines, then, as Figure 22 indicates, it should be able to descend to the Earth at the north and south magnetic poles. An appreciable amount does precisely that. It is the collision of this gas with the upper parts of the Earth's atmosphere that causes aurorae. Most of the light comes from gas particles in the atmosphere which have been knocked about by rapidly inflowing material from the solar wind; but, using the right instruments, light from the inflowing solar wind can also be seen from the Earth's surface.

Our second look at the capture of gas by the Earth therefore suggests that we must be acquiring an appreciable quantity, but this is still not the whole story. The solar wind, besides contributing material, can also remove gas as it sweeps past the Earth, peeling it off from the outermost reaches of the atmosphere. The amount taken is difficult to estimate, so it is still uncertain whether there is a net gain of material from the solar wind, or not. Cosmic rays from sources other than the Sun do accumulate on Earth, but the amount of material involved is tiny.

This chapter has shown what happens when the Earth encounters the solids and gases that surround it in space and the final question is how much material the Earth acquires in these encounters. As might be expected from the difficulties of observation, it is almost impossible to give an accurate answer, but we can estimate, very roughly, an average infall of somewhere between 5 and 50 lb (2 and 20 kg) per square mile (2.5 km^2) of the Earth each year. This is a totally negligible amount compared with the mass of the Earth.

3
Families and Relationships

We have seen that most asteroids follow a path between the orbits of Mars and Jupiter, but when the orbits of these main-belt asteroids are looked at in more detail, they show some interesting features. You might expect that the asteroids would spread out over the main-belt region in a fairly uniform way, but there are actually a number of regions in the belt where few asteroids move. The existence of these empty spaces was recognized over a century ago by the American astronomer, Daniel Kirkwood, and they are therefore usually called the *Kirkwood gaps*. To explain why they occur requires a diversion.

If you jump up and down on the end of a diving board, you soon notice that the amount the board moves depends on how rapidly you jump. If you bounce as quickly as you can, the diving board does not move too much. If you space your jumps out, you find that there is one rate of bounce at which the end of the board moves up and down by quite a long distance. Expert divers jump at this rate in order to obtain the maximum lift-off from the board. What they have found is the point at which the board resonates. The idea of *resonance* is not one which is mentioned much in ordinary life, but it is a concept which is put into practice quite commonly. For example, if you tried to make a guitar or violin out of strings attached to a flat plank of wood, the sound produced would be very weak. But attach the same strings to a properly-shaped hollow box, and the sound you make can be heard all over a large hall. The strings are vibrating at the same rate in each case, but the box is constructed to respond to the vibrations, whereas the plank is not. The resonance produced by the box greatly increases the volume of the sound.

Now resonance can also occur in astronomy. Suppose we think of two planets – say Mercury and the Earth – orbiting the Sun. It is one of the rules of the solar system that the further away an object is from the Sun, the more slowly it moves. Mercury is a good deal closer to the Sun than the Earth is, so a year on Mercury – the length of time it takes to go round the Sun once – is only about a quarter of an Earth-year. Mercury, on the inner track, keeps catching up with the Earth and passing it during the course of one Earth-year. Every time two planets pass each other they indulge in a mutual gravitational tug. Normally this pull can occur anywhere round their orbits, because the point at which they pass changes from one circuit to the next, but this need not always be true. If, for example, Mercury were to go round the Sun exactly four times for each circuit by the Earth, then the two planets would always pass at exactly the same points in their orbits each time. In this example, the Earth could be seen as the diver and Mercury as the diving board, with the Earth pulling on Mercury at just the rate to make it respond. Under these circumstances – which correspond to resonance – the actual paths of the planets can be affected because the gravitational effects can build up with time.

The basic requirement for resonance then is that the time taken for one body to circle the Sun should be a simple fraction of the time the second body takes. We can now return to the Kirkwood gaps. These are usually labelled in terms of their distances from the Sun; but each distance corresponds to a certain length of time for orbiting the Sun. The distances we have previously used for the inner and outer boundaries of the main asteroid belt correspond to periods ranging from somewhat over three Earth-years at the inner edge to about six years at the outer edge of the belt. Jupiter, lying further out, takes considerably longer for one circuit – a little less than twelve years – but this planet has far more gravitational influence on the asteroids than any other body except the Sun. An asteroid in the main belt which takes just less than four years to circle the Sun will be continually lining up with Jupiter at the same points in its orbit: it will be in resonance with Jupiter. Because Jupiter is far more massive than any asteroid, it will be totally unaffected by the consequent build up of the asteroid's gravitational pull; but, conversely, the asteroid is likely to be greatly affected by Jupiter's pull.

One of the Kirkwood gaps corresponds to a point at which an asteroid orbits the Sun three times for every one orbit by Jupiter. It follows that this gap occurs where asteroids are in resonance with Jupiter. Presumably, if you suddenly dropped an asteroid into this gap, the interaction with Jupiter would grow until the asteroid was pulled into a different orbit, which would immediately solve the problem. Having reached a different distance from the Sun, the asteroid now takes a different time to complete one circuit, so it no longer goes round the Sun exactly three times for each circuit of Jupiter. Consequently resonance no longer occurs. The other gaps can be explained in the same way. For example, another gap corresponds to the distance where an asteroid would circle the Sun exactly five times for every two circuits by Jupiter.

This is the way the Kirkwood gaps were explained to me when I started astronomy, and I still follow the same line in my own teaching. But resonance effects are actually much more complicated and interesting than this. Look again at the length of time planets take to circle the Sun. Jupiter, as we have seen, takes a bit less than twelve years to go round once. The next planet out from the Sun, Saturn, takes slightly less than thirty years. Clearly, these numbers mean that every time Jupiter completes five circuits of the Sun, Saturn is completing two orbits. So resonance can occur between the two planets. Similar relationships can be found between the orbits of other planets and in fact the solar system seems to be full of resonances. But, oddly enough, far from throwing the planets out of their orbits, these resonances seem, on the contrary, to hold the planets in place. So it is possible that Bode's law simply reflects the tendency of planets to move *into* resonant positions. In other words, the law may reflect less the way in which the planets started life than the way they have interacted with each other since their formation. This conclusion leads us to the question why planets should like resonances, whilst asteroids avoid them.

The problem is further complicated by the fact that some asteroids are known which do congregate at points of resonance. We have talked about asteroids taking some fraction of Jupiter's time to circle the Sun, but the simplest possible relationship is obviously 1:1. In this case an asteroid orbits in the same period of time as Jupiter and, consequently, follows the same orbit as the planet. Observations reveal that a few asteroids, known as the Trojan asteroids (because they are named after heroes of the battle for Troy in classical times), do circulate in the same orbit as Jupiter. They fall into two groups, both of which maintain a fixed relationship with Jupiter. One group lies ahead of the planet in its orbit, always keeping its distance from Jupiter equal to Jupiter's distance from the Sun, while the other group maintains an equal distance behind Jupiter. (Those asteroids which precede Jupiter are mainly named after attackers of Troy; those following Jupiter are called after the defenders.) The existence of these groups is again due to resonance, but here the resonance herds the asteroids together. Asteroids that stray out of the groups tend to be forced back into them.

It is not very satisfactory to have an effect that can sometimes work in one direction and sometimes in another, so big attempts have been made in recent years to understand more fully what happens when two orbiting bodies resonate. For example, one suggestion is that the final result depends on whether any collisions occur in the region of resonance. It is most unlikely that the planets will ever collide with each other, but there are so many fragments in the asteroid belt that collisions there must happen quite often. These collisions may help clear the resonance zones in the asteroid belt, whilst not affecting the planets.

In astronomical terms, the asteroid belt is heavily congested. Even a rough estimate indicates that there must have been a large number of collisions between asteroids within the lifetime of the solar system.

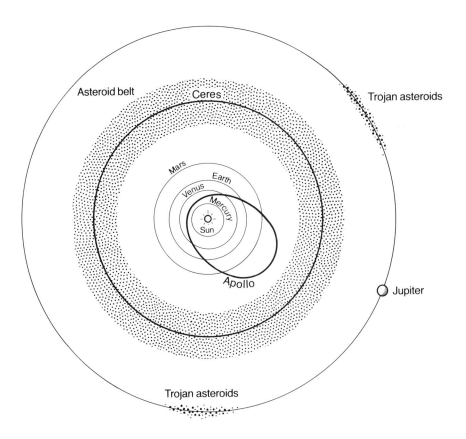

17 Although asteroids have a wide variety of orbits, three groups of particular interest are shown here. Firstly, the main-belt asteroids, with the orbit of the largest, Ceres, indicated. Secondly, the Earth-crossing asteroids, represented here by Apollo. Thirdly, the Trojan asteroids which move around the Sun in the same orbit as Jupiter – one group ahead of the planet, and the other an equal distance behind it.

Since asteroids usually come together at high speeds – maybe a few miles per second – encounters between them tend to be catastrophic, with one or both bodies breaking up into fragments. The overall result must be that the number of smaller bodies in the asteroid belt is increasing gradually with time, and this picture certainly fits well with the size distribution of fragments. In addition, there is some more direct evidence that supports the collision theory. An observer on Earth would be very lucky to see a collision, for what may occur frequently on the timescale of the solar system occurs very rarely over the span of a human lifetime. But there is additional evidence of a different kind. In some cases, several asteroids appear to be following almost identical orbits round the Sun. If the average distance of each asteroid from the Sun is compared with the angle that asteroid's orbit makes with the ecliptic, we find that there are several groups of asteroids with similar average distances and inclinations. This means that the asteroids in each group (known as an asteroid family) are following very similar orbits. Now it is usually easy to take something which is neatly arranged and disorder it, but it is much more difficult to take something chaotic and put it in good order. Consequently, several bodies in the same orbit are most unlikely to have wandered together by chance. Astronomers are much more prepared to believe that the bodies are related to each other and that this is why they follow similar paths. In the case of asteroid families, the simplest assumption is that they represent fragments of a single body which broke up; the most likely cause of such a break up would be collision with another asteroid. If we are prepared to accept this line of reasoning, asteroid families can provide us with information about collisions in the asteroid belt.

As the number of asteroids with well-determined orbits has grown, so has the number of families, but even so there are plenty of main-belt asteroids which do not belong to families. This does not of course mean that they are not collision fragments. Jupiter pulls on all asteroids, not just those which are in resonance with it, and therefore affects all orbits to some extent. This pull is supplemented by the effects of the collisions themselves. The fragments from a collision will move with different speeds, and this will put them into slightly different orbits round the Sun, so that two asteroid fragments from the same collision may gradually drift apart. In consequence, it is possible that most main-belt asteroids have been fragmented during their past careers. The exception to this rule is Ceres, which is so massive compared with any nearby asteroid that impacts would only pockmark its surface, not fragment it. With Ceres, at least, we may have a body whose existence dates back to the beginnings of the asteroid belt.

Jupiter's gravitational pull can do more than readjust orbits within the main belt. If current ideas are correct, it can lead to major changes in asteroid orbits. One of the most difficult questions in studies of asteroids is to find a mechanism for moving main-belt asteroids into planet-crossing orbits. The commonest assumption is that, since the greatest disturbance in the main belt occurs round resonance zones, it is material from these regions which is forced inwards towards the terrestrial planets. In other words, Jupiter converts circular asteroid

18 OVERLEAF Now that space probes have been dispatched to Halley's Comet, the only group of bodies in the solar system which have not been studied in detail are the asteroids. Discussions have now been under way for some time to try and plan an unmanned mission to the asteroid belt, which would look at different types of asteroid.

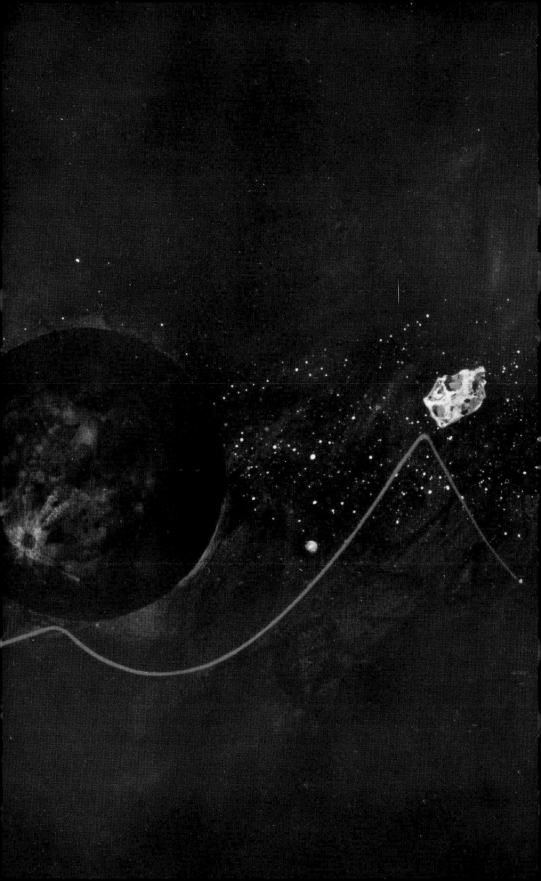

orbits into elongated ones. Even if this view is accepted, it does not solve all the problems. In 1977 a new asteroid, subsequently called Chiron, was found orbiting far out in the solar system between Saturn and Uranus. There is no obvious way in which the influence of Jupiter could put an asteroid into such an orbit.

Once asteroids have been placed in planet-crossing orbits there is the intriguing possibility of capture by planets. Besides its four large moons, Jupiter also has a number of smaller satellites, mostly a good deal further away from it. Some of these outer Jovian satellites are small rocky bodies which look suspiciously like asteroids. This suspicion is enhanced by the fact that they follow unusual paths – elongated orbits in which the satellite moves in the opposite direction to the way Jupiter spins on its axis (another example of retrograde motion). Theoretical studies suggest that Jupiter may be able to capture occasional Trojan asteroids and then release them again. If so, Jupiter is an unusual example of a planet with temporary satellites.

On the inner edge of the asteroid belt, the two small moons of Mars also look very much like asteroids and the innermost one follows a very odd orbit: it actually moves round the planet faster than Mars rotates. (So, to an observer on Mars, it would rise in the west and set in the east.) It is therefore generally supposed that the Martian moons are captured asteroids. The fact that Mars has only acquired two despite its closeness to the asteroid belt is an indication of how difficult it must be to capture asteroids. The Earth has not managed to pick up any, though it is estimated that there must be about a thousand asteroids in Earth-crossing orbits. In fact, an asteroid is more likely to collide with a planet than to be captured.

Comet orbits can be influenced by Jupiter's pull just as effectively as asteroid orbits. As we have seen, this is how short-period comets are produced. In fact, comets can approach Jupiter closer than most asteroids can because their paths cut across Jupiter's orbit. A comet discovered in 1770 actually passed between the large Jovian satellites. Some of the theories that have been postulated for asteroids can be seen in practice in the interactions of comets and planets. For example, a few comets have been temporarily captured as satellites by Jupiter during the past hundred years, though generally for fairly short periods.

We expect a cometary orbit that has been deviated by a planet to be associated in some way with that planet. Thus typical members of the Jupiter family of comets approach the orbit of Jupiter at their furthest distances from the Sun. But some short-period comets are not closely associated with any planet. For example, one comet discovered by our group at Leicester using the IRAS satellite (Infrared Astronomical Satellite) in 1983 (see Fig. 19) reaches a maximum

19 The Infrared Astronomical Satellite (IRAS), which operated during 1983, was the first satellite capable of measuring the heat emitted by celestial objects. It proved to be particularly good at detecting comets and asteroids.

distance of 14.5 AU from the Sun, where there is no planet. This suggests that the process of turning long-period comets into short-period comets is more complicated than can be explained simply by the comet having been captured by a planet in a single event. It is more likely that there are several successive changes in the cometary orbit that eventually produce the result that we see.

Of course, what can work in one direction can also work in the other. A short-period comet may have its orbit altered again by another encounter with a planet and, more often than not, this will return the comet to a longer-period orbit. In some cases, interaction with a planet can throw the comet entirely out of the solar system and it will then travel into interstellar space and be lost to us for ever. Comets, like asteroids, can also be lost by collision with planets but, unlike asteroids, they may also be destroyed if they venture too close to the Sun. In recent years, satellites observing the Sun from space have actually recorded instances of comets hitting the solar surface.

A small number of comets do not hit the Sun, but pass very close to it. Several members of this *Sun-grazing* group follow very similar orbits which suggests – as in the case of the asteroid families – that we are probably seeing fragments of one large comet which broke up. Here again, unlike asteroids, we have been able to see this activity occurring in the comets. One of the earliest examples of an observed break-up was Biela's comet in 1845–6. According to one description: 'When first detected, on November 28, it presented the appearance of a faint nebulosity, almost circular, with a slight condensation towards the centre: on December 19 it appeared somewhat elongated, and by the end of the month the comet had actually separated into two distinct nebulosities, which travelled together for more than three months.'

These *split* comets are simply more spectacular examples of the continual wastage that comets undergo. Loss of material into the head and tail must evidently lead to the final death of any comet, if it is not removed previously by one of the other hazards we have been examining. This drain on the number of comets may mean that the solar system started with a very large stock of comets (otherwise we would not still be seeing them today). Alternatively there may be some way of producing new comets, an important question that we shall put aside until the last chapter.

The loss of material from comets can have an effect on their motion. Sometimes the material from a comet nucleus, instead of evaporating gently, blows off in a large jet. When this happens the effect on the comet is a bit like the effect of a rocket on a space probe – it slightly alters the path that is being followed round the Sun. The comet is then said to be affected by *non-gravitational forces*, exemplified by the drifting apart of the pieces of a split comet.

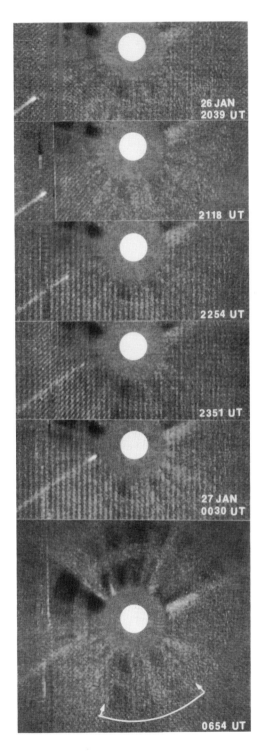

20 This sequence of photographs was taken from a satellite orbiting the Earth. A comet approaches the Sun from the lower left. It fails to reappear on the other side of the Sun (its expected reappearance position is indicated by the arc in the last photograph). Hence, it is presumed to have collided with the Sun. The rays in the background come from the Sun's corona (which the satellite was designed to observe). The times in the bottom right-hand corner are given in terms of hours and minutes of Universal Time (UT). This is equivalent to time at Greenwich, London.

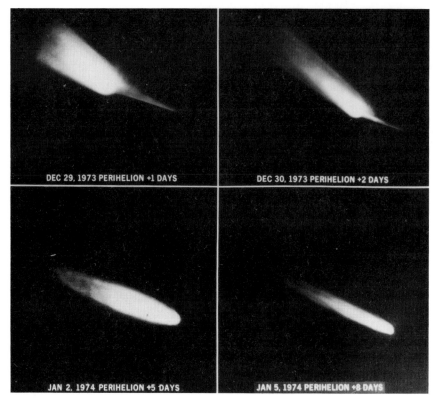

DEC 29, 1973 PERIHELION +1 DAYS

DEC 30, 1973 PERIHELION +2 DAYS

JAN 2, 1974 PERIHELION +5 DAYS

JAN 5, 1974 PERIHELION +8 DAYS

21 These are sketches of Comet Kohoutek made by astronauts circling the Earth in Skylab. The astronauts were in a better position than ground-based astronomers to observe the comet at this time. Their sketches clearly show that, from an Earth perspective, Comet Kohoutek had a pronounced anti-tail for a few days.

The sunlight pressing on a dust particle in the tail of a comet is exerting a non-gravitational force, so the orbit of the particle round the Sun is changed. This pressure is more effective on lighter particles, so if the dust is made up of particles with a range of sizes, this will lead to the different-sized particles moving in slightly different orbits. The most spectacular example of this process of separation is provided by *anti-tails*. A few comets seem not only to have dust streaming back into the tail, but also dust projecting forwards along their orbits. This is an optical illusion: both tail and anti-tail are really behind the head as viewed from the Sun. When an anti-tail appears, you are actually looking down the length of the tail from the general direction of the head. Part of the tail is sticking out on one side of the head, and part on the other side, so giving the tail/anti-tail appearance. The reason that the tail is spread out in a wide fan is that the dust particles are sorted out by their interaction with sunlight. The larger particles move away from the head at a different angle to that of the smaller particles. In consequence, if you are looking at a comet from an appropriate angle, the large particles will appear as the anti-tail, and the smaller particles as the normal tail.

Seen from a point above the orbit of a comet, the effect of the pressure of sunlight on the orbits of the dust particles is to produce a curved fan of dust, a bit like a scimitar in shape, spreading out behind the head. The gas swept back into the tail usually moves in a fairly straight line and, consequently, the dust tail and gas tail are often well separated in space. So long as you are not looking at the tails edge-on to the comet's orbit – when they appear to merge – it may well be possible to distinguish the two on a photograph.

The dust from a comet trails further and further behind the head, and is more and more dispersed as time passes. If the comet's orbit happens to cross the Earth's orbit, so that a meteor shower is produced, you can see this dispersal in action. The Earth may take days to plough through the full extent of the dust cloud, if the material was deposited in space years ago. The final step in this dispersal is obviously when the cometary dust becomes part of the general interplanetary dust. But this is not the end of the story, for the dust is still exposed to sunlight. The effect of the light streaming out from the Sun is oddly like the effect of the Earth's atmosphere on an artificial satellite. Initially, the satellite can circle the Earth in a highly elongated orbit, but, as the drag of the atmosphere takes effect, the orbit becomes more and more rounded. Finally, when the orbit is thoroughly embedded in the atmosphere, the satellite gradually spirals down to the ground. Something roughly similar happens to cometary dust. Having spread out from the original elongated orbit of the comet, it spirals down under the influence of sunlight drag towards the Sun. It usually fails to reach the Sun, evaporating into gas while still a good distance away.

The result of this complicated life pattern is that dust in the solar system is held in a form of balance. There is input from comets (especially short-period comets), and there may also be input from impacts in the asteroid belt, though some estimates of the amount of input from comets suggest that this source by itself is adequate. In addition, space probes that have ventured through the asteroid belt have not encountered major concentrations of dust there, suggesting that it is not a significant source of interplanetary dust. On the outgoing side of the balance, dust is removed by various mechanisms (for example, impact with planets), but above all by the Sun. Even this picture does not tell the whole story. The first satellite capable of measuring the heat (infrared radiation) emitted by objects in space was in operation throughout much of 1983. As we have seen, interplanetary dust reflects some sunlight to give the zodiacal light, but that part of the light which is not reflected is absorbed by the dust. This makes the dust hotter, and the satellite, IRAS, was able to detect this heat. The totally unexpected result of its observations was to show that dust is not distributed in a single band round the ecliptic, but instead occurs as three bands – one band in the ecliptic, with another above it, and the third below it. There is still no generally accepted explanation of this division, so we cannot yet be sure that we really understand the balance of dust in the solar system.

We have seen that the dust tail of a comet does not point exactly away from the Sun. The same is true, though less obviously, of the gas tail. The Sun is spinning round on its axis about once every month. This is a fairly slow rate, but, because the Sun is a large body, it still amounts to a speed of about a mile (1.6 km) per second at the Sun's equator. The result is that material leaving the Sun does not shoot straight out, but follows a spiral path, in the same way that water from a spinning lawn-sprinkler will follow a spiral through the air before it hits the ground. It follows that the solar wind does not blow out from the Sun along straight lines, but along a slight curve. The gas tail of a comet, since it is shaped by the solar wind, correspondingly lies at an angle to the line joining the comet to the Sun.

The last chapter explained that most of the gas in interplanetary space is broken up into electrically-charged particles and also showed that a magnetic body, such as the Earth, can affect the way in which charged particles move. But the discussion of the way the solar wind hit the Earth assumed that the only magnetic body present was the Earth, although, in fact, since the Sun is magnetic, so is the solar wind. Thinking in terms of the imaginary lines introduced in Chapter 2, the solar wind pulls out the magnetic lines from the Sun as it blows outwards. This means that, if we could carry a magnetic compass with us on a trip through the interplanetary gas, we would find it pointed in the plane of the ecliptic along a spiral line leading either towards, or away from the Sun.

The importance of this interplanetary magnetism becomes apparent when you compare what happens to the head of a comet with how the Earth's magnetosphere is formed. The comet, unlike the Earth, is probably not magnetic, so you would expect its interaction with the solar wind to differ from the Earth's. In fact, the differences are quite small because the magnetism of the solar wind prevents the gases in the wind – the solar plasma – from mixing with the gases from the comet – the cometary plasma. The gas in the head of the comet is pressed back and the tail is stretched out behind it into much the same sort of shape as the Earth's magnetosphere. It is like the windsock on an airfield streaming out in the wind.

Comet Burnham, discovered in 1960, was found to have a tail that wagged backwards and forwards every four days. Some years later I was reading through reports of Comet Halley's appearance in 1835. I found that the tail of this comet, too, had been observed to wag every four and a half days, and that this phenomenon occurred when it was at about the same position relative to the Sun as Comet Burnham. What was happening in both cases was that gusts in the solar wind set the tails vibrating, the rate of vibration being governed by the strength of the solar wind's magnetism at that distance from the Sun. Solar-wind gusts have their most spectacular effect on comets when they cause large knots of gas to congregate at some point in the tail. These knots are then blown further along into the tail, often making a major difference to its shape. Sometimes the entire tail of a comet blows off into space, and the head will then grow a new tail.

One important difference between comets and the Earth's magnetosphere is the way the latter can trap charged particles. As we saw in Chapter 2, solar-wind particles enter the upper atmosphere of the Earth near the magnetic poles, so producing the aurorae. Instead of entering directly, however, they are more often trapped within the inner regions of the magnetosphere for some time before they can collide with the atmosphere. They are stored near the Earth, oscillating from north to south and back again until some flurry in the solar wind shakes them down to Earth. The particles are mainly concentrated in two regions round the equator, one starting a few hundred miles above the surface of the Earth, and the other more than the diameter of the Earth away. These two concentrations, called the Earth's *radiation belts*, were discovered whilst I was a research student, and I well remember them being hailed – rightly – as the first major discovery of the space age.

When electrically-charged particles oscillate up and down, they produce radio waves: the basis of radio transmitters on Earth. Consequently, if the Earth is examined from a distance, it is found to be emitting radio noise from its radiation belts. Some thirty years ago, in the early days of radio astronomy, Jupiter was also discovered to be a source of radio noise and more detailed study showed that the noise came from a region surrounding the planet. This was located in a similar position to the Earth's radiation belts, but covered a much greater volume of space. The amount of radio noise produced was much more than came from the Earth, so it was deduced that Jupiter was far more strongly magnetic than the Earth. This conclusion has since been confirmed by direct measurements made during fly-bys of Jupiter by space probes, which have shown that Jupiter has a magnetosphere like the Earth's but much greater in extent. The tail of this magnetosphere extends downwind of Jupiter as far as Saturn's orbit.

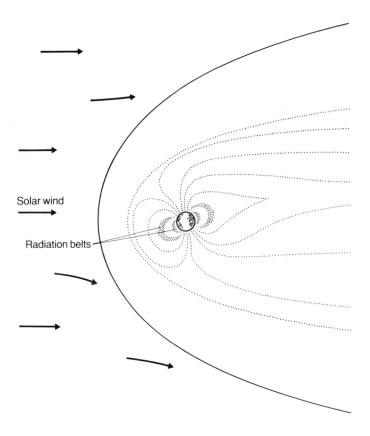

Solar wind

Radiation belts

22 *The general characteristics of the Earth's magnetosphere are shown here. The solar wind flows in from the left of the diagram. When it reaches the Earth's magnetism, a thick layer of solar gas is built up in front of this obstacle. Most of the gas slides off round the sides of the magnetosphere, stretching it out on the down-wind side to form a magnetotail (extending a long way out of the picture to the right). Some, however, is captured to produce the radiation belts round the Earth.*

Talking of the Jovian magnetosphere reminds me of an overseas research student I had at Leicester some years ago. His thesis concerned the interaction of the solar wind with Jupiter's magnetosphere, and he left his material with me to brush up for publication when he returned to his own country. Unfortunately, the government of that country was a military regime which immediately imprisoned him for alleged political activities. When he was released, he wrote and told me briefly about the ill-treatment he had received. However, that was relegated to the second paragraph of his letter, as being less important. The first, and to him much more important paragraph enquired anxiously whether his research had been accepted for publication. (I am happy to say that it had.)

Saturn, although generally similar to Jupiter in its properties, does not emit radio noise, but there is an obvious reason for this. Saturn's equator is circled by the famous rings – vast numbers of small solid fragments consisting of ice or ice-covered dust – and any charged particles that tried to oscillate from north to south and back around Saturn would need to pass straight through these rings. They would inevitably hit the material in the rings and be stopped. So Saturn cannot maintain radiation belts, although fly-bys have shown that the planet is highly magnetic. We do not know yet the extent to which Uranus and Neptune are magnetic, but it is already clear that the Earth and the major planets resemble each other in the way they interact with the solar wind. The other terrestrial planets are non-magnetic, or only slightly magnetic, and interact with the solar wind much more as comets do, without the creation of trapped particle belts.

Spacecraft measurements have shown that the solar wind actually blows throughout the whole of the planetary system and we will soon have more information about just how far it does extend. One of the Pioneer spacecraft which flew past Jupiter is now on its way out of the solar system. It has already passed beyond Pluto's orbit and has shown that the wind extends that far. Space probes are nearly always placed into orbits that are in the plane of the ecliptic. This is where the planets are, and, in any case, it takes more energy to place a space probe in any other kind of orbit. As a result, however, we have little knowledge of how the solar wind blows at appreciable heights above and below the ecliptic. Plans are now fairly well advanced for a spacecraft mission which will fly up out of the ecliptic and cross over one of the poles of the Sun, but until this takes place we must continue to rely on observations of comets which move well out of the ecliptic. Their tails indicate that the solar wind continues to blow strongly even at considerable heights above the ecliptic.

The solar wind is obviously difficult to observe directly, but the final descent of the charged particles captured by the Earth can be seen very easily in the aurorae. Because these particles come down round the magnetic poles, which are fairly close to the north and south geographic poles, you need to be at fairly high latitudes to see aurorae and the number visible each year decreases rapidly as you move to lower latitudes. When I lived in Scotland near 57°N I found that aurorae could definitely be seen more often than when I lived in southern England nearer latitude 51°N. Aurorae have the advantage that you need no instruments to observe them: they can be seen quite adequately with the unaided eye. The main requirement, as for most astronomical observations, is that you look for them when the sky is clear and dark and when you are well away from any artificial lights. It also helps to remember that the likelihood of seeing aurorae depends on how active the Sun is. The Sun's ability to pour material into space is not constant with time, reaching a maximum every eleven years or so, and then dying away again. Aurorae are more likely to be seen when solar activity is at a maximum: the next such period will be at the end of the 1980s.

Meteors are also simple things to observe. Go out on any night and wait until your eyes have become adapted to the dark and then you should see a meteor flash across the sky within a few minutes. The chances of spotting one are increased if you observe during a meteor shower, but many of these are now so spread out that the number of meteors they produce is only slightly above the background average. Fortunately, several showers occur during the autumn and winter months, when nights are long, so that there are more opportunities to go out and look for one. If you want to make your observing less random, the dates of meteor showers are sometimes published in the more up-market daily papers, or can be found in amateur astronomy handbooks.

Comets are less simple to observe. Some become bright enough to be seen with the unaided eye, but even so they are rarely up to most people's expectations. Photographs of comets can lead you to expect a spectacular sight, but you are more likely to see an undistinguished smudge of light. The first bright comet I saw was Arend-Roland (called after its co-discoverers) in 1957. There have only been a few comets to match it since, yet even it required a very dark, clear night to look at all interesting. Similarly, Comet IRAS-Araki-Alcock in 1983 came closer to the Earth than all but a few comets in recorded history, yet many of my non-astronomical friends had difficulty finding it in the sky.

Because comets are extended objects (unlike asteroids, which can only be seen as points of light), they are best viewed through instruments which cover a fair region of the sky: a pair of binoculars is often more useful than a small telescope. There are a number of amateur astronomers who do invaluable work by spending long hours searching for comets, sometimes with nothing more sophisticated than a pair of binoculars to help them and often taking hundreds of hours for each discovery. If you're interested in the photographic angle, a small telescope with a camera attached can often produce quite satisfactory photographs of bright comets.

There is a problem in finding out where to look for comets. Short-period comets follow fairly well-known orbits, so it is possible to predict where they will be for a reasonable length of time ahead, but unfortunately only a few of them are bright, so this does not really help the casual observer. Comet Halley, which we will look at in Chapter 6, is an exception, for although it is a short-period comet, it can sometimes become very bright. But often the long-period comets are the more spectacular and the position of these is much more difficult to predict. If they become bright enough, of course, their progress is likely to be followed in the media and their position given in the newspapers.

No asteroid can be seen with the unaided eye. The larger asteroids can be observed fairly readily through a small telescope, so long as it can be pointed reasonably accurately, and there are a number of handbooks which give the positions of the brighter asteroids for those who are interested. (They appear, for example, in the *Astronomical Ephemeris*, which is published annually, and can be found in any major reference library.) To observe faint asteroids or comets requires large telescopes, so work on them is mainly restricted to professional astronomers.

Although, as we shall see, meteorites are of major importance in understanding the history of the solar system, few people ever set out to observe them and most pass their lives without ever knowingly being close to one. Oddly enough, it seems to be the case that the harder it is to observe objects of the solar system in motion, the more information about the solar system they seem to provide. If I had to assess the relative importance of each object for our understanding of solar-system history, I think I would put meteorites first, asteroids and comets second, with meteors and aurorae last.

4
Change and Decay

The basic problem for astronomers is that they can do no more than observe the objects they are interested in: they cannot experiment with them. As an astonomer you learn to be grateful for whatever information the object you are studying is prepared to yield; more than most scientists, astronomers are detectives always seeking new clues. One way of expanding the amount of information available is to look at as many different sorts of radiation from objects of interest as possible, but the main difficulty in following this line is that the Earth's atmosphere blocks out many types of radiation. It allows through only ordinary light and radio radiation, plus a little bit of ultraviolet and infrared (i.e. heat) radiation. As their names imply, *ultraviolet* is the radiation which starts beyond the violet end of the visible spectrum and *infrared* starts beyond the other – the red – end of the spectrum. This means that, if we could see all the radiation from a rainbow, we would find that both the top and bottom edges extended out further. Different animals are sensitive to different parts of the spectrum: for example, some insects can see in the ultraviolet. This allows them to distinguish between flowers which, to us, have the same colour.

This problem with the Earth's atmosphere has been overcome, in part, by using artificial satellites suitably equipped with instruments, which circle the Earth well above its absorbing atmosphere. Unfortunately, however, appropriate satellites are not always available. For example, there has only ever been one satellite for observing heat radiation from space – IRAS – and it is no longer operating (see Figure 19).

What we are concerned with in astronomy is *remote sensing*. This term is usually restricted to meaning looking at the surface of a planet, especially the Earth, from a fairly short distance. But the problems are the same if the technique is reversed. In one case, you have to look into space through the obstructing veil of the Earth's atmosphere; in the other, you look back to Earth through the same veil. In both cases, you try to interpret what you can see in terms of basic laws and data established in the laboratory. For remote sensing of the Earth, however, it is usually possible to cross-check your conclusions, to go out into the field, for example, and see whether your deductions about the rocks in a given area are correct. This important stage is not usually feasible in astronomy, though in the study of space garbage you can take a step or two in the right direction. We have not yet landed on an asteroid, or hit a cometary nucleus with a hammer (though plans to sample both with space probes are currently under discussion). On the other hand, as we saw in Chapter 2, material from both does seem to be reaching the Earth. What we cannot do is to assume immediately that things received on Earth are the same as they would be if they were in space. We may, for example, be getting a very biased sample, which has undergone various selective changes in the process of reaching Earth. So cross-checking is a good deal more complicated than for the remote sensing of terrestrial surfaces. Still, people with interests in space garbage are obviously ahead of most astronomers.

There is one important point about remote sensing which applies as much to the Earth as to other objects of interest to astronomers. What we are examining is our world as it is today. What we would like to know, however, is how all the objects that interest us first formed, and whether they have changed drastically in the years that have passed since their formation. So one of the great questions in astronomy is whether we can determine the history of an object from the way it is today.

Change with time is part of the general problem of evolution. In biology, one of the basic evolutionary beliefs is that succeeding generations of plants or animals may differ slightly in their characteristics, but that changes in the population as a whole occur sufficiently slowly for relationships between differently evolving groups to be identified. If we assert, for example, that cows have more characteristics in common with sheep than with insects, then we tend to think we are saying something significant about past evolutionary links between these different groups. Cows and sheep are more closely related to each other in their origins than they are to insects. In biology, the first step in exploring these relationships has always been to place the animals or plants in a classification scheme based on their observable characteristics. If we want to explore the question of origins in astronomy, then we must adopt the same approach. When we look at the asteroid belt we see a large number of objects with differing characteristics. One of the questions that has been looked at in some detail in recent years is whether we can classify the asteroids in such a way as to provide information on possible relationships between different groups. The first step in pursuing this approach is to decide what characteristics we should use in putting together the classification.

Asteroids are visible purely because they reflect sunlight. One of the characteristics of an asteroid surface that we can investigate is what the reflected sunlight tells us about its composition and colour. It is obvious that not all materials reflect light equally well – a dark material, such as soot, for example, reflects much less light than a material like snow. In the case of asteroids only a few are reasonably good reflectors, while very few reflect more than a quarter of the light that falls on them. One large group of asteroids separates out immediately as being particularly dark – on a par with soot – reflecting 5 per cent or less of the sunlight.

When considering colour, the important property is not how well the asteroid reflects, but whether the reflected light differs in colour from the original sunlight. The group of poor reflectors just described does not change the colour of the reflected light very much, but there is a large group amongst the better reflectors which reddens the sunlight appreciably on reflection.

These two simple measurements – the amount of sunlight reflected and its colour – actually isolate what are generally agreed to be the two main types of asteroid. The poor reflectors are called C-type asteroids, whilst the reddish-coloured ones are labelled S-type asteroids. These names stem from an attempt to guess at the kinds of material which would produce these characteristics: asteroids designated as C-type may have carbonaceous material on their surfaces, whilst S-types may have mainly silicate material (i.e. much like rocks on Earth).

A closer look at the colour and amount of reflected light reveals additional possible types of asteroid. For example, M-type asteroids have been distinguished which reflect about as much sunlight as S-types, but are less reddish. (The 'M' stands for metal, since the way this group reflects light may be due to metallic material on the asteroid surfaces.) This search for asteroid groupings has not been entirely guided by the observational data, for it has been argued from the start that, if meteorites are really asteroidal fragments, then they can be used as a guide in any classification. We should be able to measure meteorite characteristics – such as colour – in the laboratory, and, by comparing the results with observations of asteroids, work out relationships between the two. This process is equivalent to what is done in the remote sensing of rocks on the Earth's surface: the difference is that a number of assumptions have to be made for asteroids that are not necessary for the Earth. One obvious assumption is that meteorites come from asteroids, whereas in fact some may come from dead comets, and so be unrelated to main-belt asteroids.

Even if most meteorites do come ultimately from the asteroid belt, it does not follow that they provide a proper sample of the material in the belt. As we discussed in the last chapter, the shift of main-belt asteroids into Earth-crossing orbits probably requires special circumstances. This can mean that we, on Earth, acquire only a restricted diet from the full asteroid menu. If you examine the distribution of different types of asteroid across the main belt, you will find that they are by no means uniformly distributed; the M-type asteroids, for example, appearing to congregate near the Kirkwood gaps in the middle of the belt. If transfer to Earth-crossing orbits can only occur near resonances with Jupiter, this could mean that we receive a disproportionate amount of M-type material on Earth. Equally, there could be main-belt material that we do not sample at all. Some asteroids at the outermost edge of the main belt (and some Trojan asteroids) have been grouped together as an RD-type. The initials, in this case, simply describe the appearance of the asteroid (**r**eddish and **d**ark), rather than guess at the materials present on its surface. The reason is that we do not know whether there is a meteorite analogue on Earth for this group.

Another problem that faces attempts to compare observations of meteorites in the laboratory with observations of asteroids in space is the different conditions under which the two are being measured. For example, asteroid surfaces are exposed to the near vacuum of space, whereas meteorites are sitting at the bottom of the Earth's atmosphere. A meteorite can, and probably will, absorb gases from the Earth's atmosphere, so possibly changing the way it reflects light. There is also the question of the physical state of the material on an asteroid surface. A meteorite is a lump of solid rock, but the surface of a typical asteroid, as we shall see in the next chapter, is more likely to be covered by small grains of rock, like sand on the seashore, than by large extended rock layers. How a material reflects sunlight can be affected by the size of the pieces in which it occurs.

One result of all these problems is that we are not totally certain which meteorites resemble which asteroids. As is also true in the world of plants and animals, it is sometimes the unusual – the freak – which can provide most illumination. One approach, therefore, is to look for asteroids with unusual properties, and then see whether there are any meteorites with similar unusual properties. The most interesting example from this point of view is the asteroid Vesta, which will be discussed in detail in Chapter 6, but another example is the very small group of E-type asteroids. These have unusual properties which resemble the characteristics of a small and unusual group of meteorites called *enstatite achondrites*. ('Achondrites' will be explained later in this chapter.) 'Enstatite' is the name given to the magnesium silicate which makes up a good proportion of these meteorites, and 'E-type' is derived from enstatite. Enstatite achondrites have sufficiently unusual light-reflection properties for the link with the corresponding asteroids to be reasonably likely, despite all the measurement problems.

Most of the known asteroids fall into one of the main groupings, but there are a significant number which cannot be fitted into the classification. These are denoted as U-type – meaning unusual, and also unclassifiable in the main groups. But one obvious feature of classifying asteroids is that the more information you have about them, the more distinct groups you can find. This is also true of biological classication: given sufficient information, each plant and animal forms a group of its own. Biologists have attempted to handle this problem by feeding all the characteristics of the plants or animals concerned into a computer, and then asking the computer to sort them out in sequence according to how similar the groups are when compared over all their characteristics. What you get out of the computer as a result looks a bit like a family tree. Things that are greatly different from each other branch off at the top of the diagram, things which have more similarity branch off at lower levels, until, at the bottom of the diagram, the branching corresponds to things that only differ in minor details. For example, insects would separate out from cows and sheep at the top of the diagram; sheep and cows would divide at a lower level.

At Leicester, there is a microbiology group applying computers to the classification of bacteria. We have applied their methods and computer programs to the classification of well-observed asteroids and the resulting diagram shows that the main division is between C-type and S-type asteroids. This occurs at a high level, whilst other divisions come in lower down. One useful aspect of this method is that it allows us to look again at the U-type asteroids, so that we can say a little more about them. For example, the second-largest asteroid, Pallas, is labelled U-type. In our diagram, however, it can be seen to have more properties in common with C-type asteroids than with S-type asteroids (and, indeed, attempts have been made recently to classify Pallas into a new intermediate group).

As we noted in Chapter 1, the most informative way of examining the sunlight reflected from asteroids is to look at its spectrum in detail. Less detailed measurements of colour give you some idea of the materials present, but pinning down the specific minerals normally requires a spectrum. What you are looking for is indications of lower reflectivity over specific restricted regions of the spectrum. By comparing the way various minerals reflect light in the laboratory, you can begin to work out which mineral produces which depression in the spectrum. For example, several asteroids show a marked depression in their spectrum in the near infrared (that is, in the sunlight reflected just beyond the red end of the spectrum, which can only be detected with special instruments), and it has been found that this depression in the spectrum is probably caused by the silicate minerals that are well known in terrestrial rocks.

One of the most interesting discoveries from this kind of work has been the observation that water is present in the surface materials of a few asteroids. This does not mean that there are stretches of liquid water on these asteroids, for, since they have no atmosphere, any such pool would disappear rapidly. But quite a number of minerals can capture water, combining with it chemically so that it actually forms part of the mineral, and it is water in this form that has been observed. The initial observations revealing this feature were made by US astronomers, who concentrated on C-type asteroids, so it seemed to us at Leicester to be worth checking whether any S- or M-type asteroid showed the same feature. In order to do this we organized an expedition to UKIRT (United Kingdom Infrared Telescope) to look at the spectra of some asteroids in these groups.

UKIRT is in Hawaii, where it has been built at a height of some 13,000 ft (4000 m) on the extinct (we hope) volcano of Mauna Kea in order to get above as much of the Earth's atmosphere as possible. From this altitude, the instruments can peek a little further into the infrared than at ground level. It is very pleasant down at the observatory offices in the small town of Hilo on the coast and the residential accommodation at a bit less than 9000 ft (3000 m) up the mountain is also acceptable enough. But the actual high-altitude observing can sometimes induce quite unpleasant symptoms and I always seem to acquire a permanent headache at that kind of height. My chief consolation comes from the advice my doctor gave me: he told me that younger people were likely to be even more affected than geriatric cases like myself.

The observational data we brought back from this trip proved quite difficult to interpret. Bodies in the solar system reflect some of the sunlight falling on them and absorb the remainder, but this absorbed light is then re-emitted as heat (i.e. infrared radiation). There is an overlap region in the near infrared where some of the radiation you measure from asteroids is reflected solar heat and some is heat emitted by the asteroid itself. Disentangling the two components can be tricky, but it has to be done if you want to look for depressions in the reflected spectrum. This was particularly important in our case because the characteristic dip indicating the presence of water falls in this overlap region. It took us some time to work out the results, but eventually we managed to show that none of the S- or M-type asteroids we observed appeared to possess water.

The significance of this work is that some of the carbonaceous meteorites, but not meteorites of other types, show signs of having been extensively affected by water in times past. They still possess some water bound in their minerals today; so the identification of bound water on some C-type asteroids, but not on other types of asteroid, fits in well with the meteoritic evidence.

In fact, the US observers have suggested an even more exciting possibility. Ceres is the largest and brightest of the asteroids, and it is feasible to examine its spectrum in considerable detail. This seems to indicate an extra dip in the near infrared spectrum due to free water, rather than to water bound into minerals. As we have seen, such free water cannot exist for long on Ceres because it would evaporate into space. For free water to be there today, there must be a constant source of water welling up from the interior. If that source has been there throughout the whole career of Ceres, then it must have started life as a very wet lump of rock.

We have talked mainly about asteroid spectra in the infrared: this is because several of the common minerals in rocks tend to produce depressions in this region of the spectrum. However, minerals can also affect other parts of the spectrum and one of the things I have been concerned with for some years past is whether the materials present on asteroid surfaces can be pinned down by observations in the ultraviolet part of the spectrum. Since ultraviolet radiation is absorbed high in the Earth's atmosphere, this question requires an appropriate satellite if it is to be answered. Fortunately, an excellent satellite – IUE (International Ultraviolet Explorer) – has been available for this purpose during the past few years and we have been able to observe with it from a ground station in Spain. The instrumentation in the satellite can be instructed to look at an object and produce observational data in much the same way as a computerized ground-based telescope. The results we have derived from ultraviolet observations suggest that several asteroids are covered with minerals containing elements like iron and titanium.

In general terms, it is easier to obtain information about solids by observing in the red and infrared, but about gases by observing in the blue and ultraviolet. Since comets contain both dust and gas, they need to be examined throughout the spectrum. However, the infrared spectrum of the dust is harder to interpret than infrared spectra of asteroids because, in this case, the results depend on the size of the dust grains as well as their composition. Nevertheless, it seems reasonably clear that cometary dust contains both silicates and metals – much like the mixture found in meteorites. When comets approach the Sun closely, some of the dust evaporates, and then metals, such as sodium, appear in the head of the comet as gases. A whole range of normally gaseous materials have also been observed. These may be electrically neutral in the head, but have usually become electrically charged by the time they have been pushed back into the tail. In neither case are we seeing the original constituents released from the cometary nucleus: the effect of sunlight has rapidly broken these down into the simpler materials we see. What we can do is to attempt to act like detectives again and to deduce what was in the nucleus from observing what appears in the head and tail. The attempts to answer this question have led in recent years to a belief that the original nucleus is usually made up of water together with carbon dioxide and/or carbon monoxide and smaller amounts of rather more complex materials (such as the highly poisonous gas hydrogen cyanide).

These constituents explain why the nucleus is often referred to as a *dirty snowball*: the water and other frozen gases provide the 'snow' and the dust provides the 'dirt'. However, the snow is more complex than just a mix of frozen gases and it seems that the water can actually form compounds with some of the other gases present, much as it does with minerals on the surfaces of some asteroids. If the various frozen gases were just mixed together, you would expect them to boil off in sequence as a comet approaches the Sun. The higher the melting point of the material, the nearer the Sun the comet would have to be for that material to evaporate. But, in many comets, most of the gases start to appear together at about 3 AU from the Sun. The significance of this distance is that it corresponds to the point where you would expect frozen water to begin evaporating rapidly. Evidently what happens is that, because the gases have formed compounds with the water, they cannot evaporate extensively until the water itself evaporates and releases them. On this interpretation the development of a comet is governed by the water it contains.

All the gases in the nucleus, including water, are readily broken up by the radiation from the Sun. The break up of water, in particular, produces great quantities of hydrogen gas which has a spectrum that is strongest in the ultraviolet. One of the first discoveries made with satellites observing in the ultraviolet was that visible comets are typically surrounded by a huge cloud of hydrogen gas and in this respect comets resemble the solar wind into which their gas merges. Thus an analysis of the spectrum of aurorae shows that most of the light comes from gases in the Earth's upper atmosphere – oxygen and nitrogen – as they are knocked about by the solar-wind particles; but a closer examination indicates that the inflowing wind particles also give out light, and that this light is typical of hydrogen. The predominance of hydrogen is hardly surprising: after all, the wind emanates from the Sun which is, itself, predominantly made of hydrogen.

Like the spectrum of aurorae, meteor spectra include a lot of light from the heated atmosphere, but even so some indication can be found of what the meteor itself contains. The problem with meteor spectra is the same as with meteor photography – catching them in the act. The obvious meteors to look for are those that come in showers, since they are, to some extent, predictable. The differences between shower meteors, judging by their spectra, have a similar range to the differences between meteorites, and this has been confirmed by laboratory analysis of interplanetary dust particles collected in the upper atmosphere. Whatever frozen ices may have been present originally, they do not show up readily in these studies.

Of all the bodies discussed in this book, we certainly know most about meteorites. We can handle them in the laboratory and apply to them all the methods developed for studying terrestrial rocks. As one of my friends once said to me, the nearest thing in appearance to a meteorite is a lump of concrete with a coating of tar on one face. The black tar represents the fusion crust of the fragment (produced, as we saw in Chapter 2, by passage through the atmosphere), while the grey aggregate of the concrete closely resembles a section through a stony meteorite: you have to look closely to see the difference. The best give-away that you are dealing with a stony meteorite is usually the presence of shiny grains of metal distributed at random through the stone. If you examine the meteorite in more detail you will often see that there are small balls – a bit like undersized marbles, but not always spherical – embedded in the rock. These are called *chondrules*. They are more difficult to distinguish than the metallic grains because they consist of silicates like the surrounding rock. However,

23 A cross-section of a meteorite seen through a microscope. There is a chondrule with a fairly circular cross-section in the centre of the picture, with a smaller, rather more distorted companion on the right.

the larger chondrules can be detected with the unaided eye: a piece of the Barwell meteorite had a particularly large chondrule looking like a small pebble sticking out of one of its faces. But many chondrules are smaller – the size of a pinhead – and it is helpful to use a small hand lens in looking for these.

Nothing like chondrules is found in terrestrial rocks. Chondrules clearly formed before the meteorites themselves, and somehow became embedded in the rock as it came together. If you examine them in detail you will see that their interiors are glassy. For this glass to have been formed, silicate rocks must have been heated above their melting points and then cooled down very quickly. We can suppose that chondrules started life as molten droplets, which froze rapidly into glassy spheres. There has been a great deal of speculation about how they were produced. One possibility is that they resulted from impacts – in much the same way that the tektites we will discuss in the next chapter are supposed to have been produced by impact on the Earth.

Stony meteorites that contain chondrules are called *chondrites*. Besides these chondrules and the metallic grains (which consist of an iron/nickel mixture), stony meteorites differ from terrestrial rocks in containing appreciable amounts of iron sulphide (i.e. a combination of iron and sulphur). This is usually taken as an indication that the meteorites formed in a place where there was no oxygen, since the sulphide interacts with oxygen. The remaining minerals found in meteorites are generally much the same as common minerals in terrestrial rocks.

All the types of meteorite discussed in Chapter 2 – irons, stony irons and stones – can be subdivided further. Stony meteorites, in particular, can differ appreciably from one another in their characteristics. Various ways of classifying them have been tried down the years, with the unfortunate result that the same meteorite may be grouped in different ways, depending on the classification scheme you are using. The commonest method of subdivision makes use of the fact that even those meteorites classified as stones still contain an appreciable amount of iron (though obviously a good deal less than is found in stony irons). So it is possible, for example, to talk about the *high-iron* group of stony meteorites. This form of classification is obviously based on the composition of the stones, but meteorites can also be classified according to the crystals of the various minerals they contain and whether these are clearly defined, or tend to merge together. A classification in these terms cuts across classification by composition. For example, a meteorite with a lot of iron may have either well-defined, or ill-defined crystals.

The significance of these two types of classification, and the reason they can clash, is because they measure different properties of the meteorite. The amount of metal present relates to the composition of the rock, established when the meteorite first formed, while the texture of the crystals depends on the amount of heating the rock has received since it formed. If a rock is heated at a temperature below its melting point, the crystals tend to become more granular in appearance, and individual features (such as chondrules) become less distinct. The two methods of classification together tell us that stony meteorites must have had more than one place of formation and more than one history of subsequent development.

As mentioned in Chapter 2, the carbonaceous meteorites have a strikingly different composition from the remainder, not only because they include carbon compounds, but also because of the almost total absence of iron/nickel. On examining the minerals present in carbonaceous chondrites, it becomes evident that the absence of heavy metal reflects a general scarcity of materials formed at high temperatures. The overwhelming bulk of these meteorites is made up of low-temperature materials and, in particular, they are the only group of meteorites to contain large amounts of water-bearing minerals. For example, Epsom salts, which dissolve readily in water, sometimes occur in carbonaceous chondrites as long thin veins of material running through the meteorite. The only obvious way of depositing material in this form is to have cracks in the rock penetrated by liquid water. This subsequently evaporates, leaving the previously dissolved salt behind. As we have seen, signs of water provide a fascinating tie-in with some C-type asteroids.

The other special interest of carbonaceous chondrites, apart from the evidence for water, lies in the nature of some of the carbonaceous material present. It contains small quantities of chemicals which, on Earth, are associated with living processes. Such basic compounds as amino acids, which are involved in building protein in our bodies, have been found in recent years in some carbonaceous chondrites. When I first became interested in meteorites, very little material from carbonaceous chondrites was available for study, but since then there have been a few big falls of carbonaceous meteorites. One of these was in 1969 at Murchison in Australia. It was material from this fall which provided the first evidence for amino acids in meteorites and it is virtually certain that these substances were not created by life processes, but were produced naturally. In fact, the study of carbonaceous meteorites may tell us something not only about the early days of the solar system, but also about the origin of life.

Carbonaceous chondrites can never have been greatly heated since their formation because otherwise both the water and the amino acids would have disappeared. This is one reason why they are often referred to as the most 'primitive' meteorites – meaning the least changed since their birth. But there is another reason for this name, the explanation of which takes us on a slight diversion.

We saw in the first chapter that the Sun and Jupiter are made up of very similar materials – mostly hydrogen and helium. The other major planets diverge to some extent from this composition, but the terrestrial planets are totally different. Suppose we had very large hands indeed, and that we plucked handfuls of material from our local region of the Universe at random. Will different handfuls tend to have the same composition, or will the materials present change radically from one handful to the next? The answer to this question has been built up over many years of astronomical observation and it appears that, unless you happen to light on some most unusual part of the Universe, all the handfuls will have generally similar compositions. Moreover, that composition will be much the same as the composition of the Sun and Jupiter, with hydrogen and helium by far the commonest elements. The next most common materials will be such elements as carbon, nitrogen and oxygen, which make up a good portion of our own bodies. All the other elements will be much rarer, with iron as a slight exception. In other words, the Universe around us has a recognizably similar composition – usually referred to as the *cosmic abundance scale* – which is typified in the solar system by the Sun and Jupiter. The Earth and the other terrestrial planets deviate widely from this scale, and so do most meteorites. Carbonaceous meteorites, on the other hand, mimic cosmic abundances moderately well. They are obviously deficient in hydrogen and helium, which, as gases, have little place in rocks, and the carbon, nitrogen and oxygen which they contain have also been a little affected since, though they often form solid substances, they can also combine to produce gases. For example, oxygen is a major constituent of most rocks, but it can also be incorporated into substances such as water, which evaporate easily. These qualifications apart, the carbonaceous chondrites seem fairly unchanged in composition from average material outside the solar system, which is another reason for calling them primitive.

The chondrites – that is, the majority of stony meteorites – are typically made up of lumps of material which formed separately and became welded together, but without totally changing the original properties of the lumps. Most other meteorites show signs of having been melted during their past history. This is even true of one group of stony meteorites – the achondrites. These meteorites contain no chondrules and have mostly been molten at some stage in the past – like lava from volcanoes which has solidified. In the stony-iron meteorites, it is also clear that molten metal mixed with the silicates. Finally, like the achondrites, most of the iron meteorites seem to have been molten in their previous history, but they must have cooled slowly to crystallize in their present form because the crystals of iron/nickel are very large. If solidification takes place rapidly, there is only time for small crystals to appear.

Iron meteorites, like stony ones, can throw light on the conditions in which they started life. Mixed in with the iron there are several more exotic elements, one of which is germanium, mostly known on Earth in connection with its use in transistor radios. This and other rare elements can be used as tracers to compare the nature of the different iron meteorites. An analysis of a wide range of iron meteorites suggests that they fall into twelve groups, each of which originated in a different place.

You can see by now that a prime factor in the history of most meteorite groups is the way they have been affected by heat. The question of how rapidly meteorites cooled down is therefore important because it reflects on their subsequent development and it also gives us some clues as to their origin. The more deeply buried material is, the longer it takes to lose heat. Almost all types of overlying material act as an insulating blanket, like the insulating material you wrap round a hot-water tank. This is the reason why the core of the Earth remains hot today; it is simply because it is so far below the Earth's surface. So an estimate of the rate at which a meteorite cooled can give us an indication of the size of the body from which the meteorite comes.

There are various ways of estimating the rates at which meteorites cooled, but the most interesting depends on the existence of a special type of radioactivity. Describing it involves us in another diversion. In Chapter 2 we talked about the radioactivity that is created in meteorites by cosmic rays, but meteorites also have their own intrinsic radioactivity, which may vary both from meteorite to meteorite and at different places within a single meteorite. The various radioactive elements break down at different rates, but, for a given type of disintegration, the rate is always constant regardless of the local conditions (e.g. whether the rock is at a high or low temperature). How fast radioactive materials break down is usually measured in terms of their *half-life*. This is the length of time required for half of the radioactive element concerned to disintegrate into something else. For example, if you started with a pound weight of radioactive material that had a half-life of a million years, then after this length of time had elapsed you would be left with half a pound, after another million years you would have a quarter of a pound, and so on.

Radioactive materials can be divided into two groups: those with half-lives a good deal less than the age of the solar system and those with half-lives of the same general order as the age of the solar system. Unless the former group has been recreated in some way, none of its members will be around today, as virtually all such material will have decayed away. But the existence of radioactive material in the latter group enables us to date the beginning of objects in the solar system.

Although radioactive materials with a short half-life have now disappeared from naturally occurring rocks (unless regenerated by cosmic rays), we can at least try and look for 'fossil' traces of their presence. This has been labelled *the search for extinct radioactivity.* One of the most interesting extinct elements to look for is plutonium – a material manufactured in our own lifetime for peaceful or warlike uses of nuclear energy. There was a fair amount of plutonium about in the early solar system, but as its half-life is only 82 million years (much less than the age of the solar system), if you want any today, you have to manufacture it. Most radioactive materials break down via a series of small steps until they are finally converted into an element that is stable (i.e. does not break down further). For example, uranium passes through a number of radioactive stages, emitting electrically-charged particles on the way, until it is converted into lead. Since lead is non-radioactive, the process finishes at this point. In some cases, however, the jump from radioactive to stable is taken in one large step, instead of several smaller ones. The process by which this occurs is called fission: typically the radioactive element splits into two much simpler and stable elements. Plutonium is an element that undergoes fission.

Imagine then that you have some plutonium sitting in the middle of a rock. Fission occurs, and the two new particles created shoot out in opposite directions at high speed. Each excavates a tiny tunnel through the rock before being brought to rest. The tunnels are too small to be seen readily, but, if you slice up the rock and soak it in appropriate chemicals, you can enlarge the tunnels by chemical attack. Put the slice of rock under a microscope and you can see the enlarged tunnels quite easily. I played about with this method when it was first proposed, and was astonished at how readily the tracks could be distinguished. It is fascinating to be able to see radioactive processes at work so directly.

Now comes the crucial point that leads us back from this diversion to our original concern with heat. Though the rate of decay of plutonium is not affected by the temperature of the rock, whether rocks record tracks will depend on how much they are heated. Above a certain temperature (which varies from rock to rock), the tunnels created by plutonium fission tend to close up again after the event. If all the fission took place when the rock was hot, use of chemicals now would reveal no tracks. Put another way, the longer it took the rock to cool down to a temperature where it can retain tracks, the fewer tracks you will see, simply because more plutonium had time to disappear before the recording process started. So a count of the number of fission tracks present gives some guide to the rate at which the rock

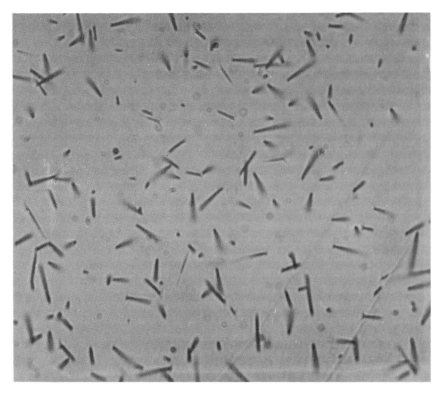

24 *When a heavy element, such as plutonium, breaks up spontaneously, the fragments are ejected at high speeds, and plough for some distance through any surrounding material. These tracks can be treated chemically to make them bigger. They can then be seen through a microscope, as in this picture.*

cooled. Moreover, as the different minerals present in a meteorite begin to retain signs of plutonium fission at different temperatures, the cooling rate of the meteorite as a whole can be found by comparing the number of tracks in the various minerals. These measurements suggest that meteorites cooled at a rate of about 10°C (20°F) every million years – some a bit faster, some more slowly. This rate of cooling could not be reached unless meteorites were never buried more than 100 miles (150 km) below the surface of the body where they formed, because otherwise the insulating effect of the overlying material would have made them cool more slowly. Turning this finding round, the minimum size of body in which meteorites could have formed was 200 miles (300 km) across or more (otherwise they would have cooled too rapidly).

You may have noticed that these measurements tell us something even more fundamental about meteorites. For the tracks to appear at all, the meteorites must have reached much their present state when there was still a fair amount of plutonium about. If our understanding of plutonium is right, this must mean that meteorites formed way back at the start of the solar system, so if we can attach an age to meteorites, we can also date the origin of the solar system.

Many meteorites contain small quantities of the element uranium. Everybody knows that uranium is radioactive because its properties became common knowledge as a result of its incorporation into the first atomic bombs. In nature it breaks down very slowly, the commonest form of uranium having a half-life of 4500 million years. Since there is still a fair amount of uranium in meteorites, they cannot be vastly older than this length of time: to put an exact figure on their age you have to compare the amount of uranium present with the amount of the substance into which it decays – lead. If you assume that all the lead present in the meteorite comes from uranium, then the ratio of the amounts of the two elements gives the age of the meteorite directly. For example, if the amount of lead is equal to the amount of uranium, this means that half the uranium has had time to break down into lead. The half-life of uranium then tells us that the meteorite must be precisely 4500 million years old.

The dubious part of this argument is the assumption that all the lead present comes from uranium. Stated in this simple form it is certainly wrong. Fortunately, there are other long-lived radioactive elements present in meteorites apart from uranium and, by intercomparing the results from different radioactive processes, it is possible to arrive at a fairly good estimate of meteorite ages. We find that most meteorites were formed at about the same time; oddly enough, at a time corresponding very nearly to the half-life for the decay of uranium. So the average age usually quoted for meteorites is 4550 million years.

If you pick up a rock containing radioactive elements on the Earth's surface and carry out the same kind of dating process, you will get a quite different result. Terrestrial rocks are typically much younger than meteorites – perhaps a tenth of their age or so. This is not too surprising. The Earth has the most rapidly changing surface of any terrestrial planet. Rocks are eroded away by wind and water, deposited as sediments, heated, lifted up to form new rocks and so continually recycled. Each time the material is handled roughly some at least of its radioactive clocks will be reset to zero. The recycling process is said to be due to the Earth's tectonic activity – a word some of my students find difficult to remember. One of them wrote in an essay some time ago: 'The Earth's surface is continually changing due to teutonic activity.' The essay was returned with the inevitable comment: 'Interesting but not germane.' Even the oldest known terrestrial rocks, at nearly 4000 million years, are appreciably younger than the average meteorite. It is worth looking at this difference briefly because it tells us something more about meteorites.

When radioactive materials break down they produce heat. (It is this heat which is used in nuclear power stations to produce electricity.) When talking about cooling rates, we saw that material buried below the surface of a planet loses its heat rather slowly. If the material is radioactive, the heat it produces therefore takes considerable time to percolate out and, whilst doing so, it heats the surrounding rocks. In the early days of the solar system there was obviously much more radioactive material about than there is at present. Not only was there a greater quantity of elements such as uranium, but radioactive elements now extinct, such as plutonium, still survived. This means that the Earth must have been more greatly heated by radioactivity in its early days. It seems possible that the surface of the Earth was very hot, perhaps molten, and this explains why we can find no rocks that solidified earlier than 4000 million years ago. The fact that meteorites settled down before the Earth suggests again that they must have originated in smaller bodies that cooled down more quickly.

Over the past ten or fifteen years, there have been extensive studies of all the radioactive elements to be found in terrestrial rocks, meteorites and returned samples of Moon rocks. The result has been to harden the belief that meteorites formed very early in the history of the solar system, and that the planets – as represented by the Earth and the Moon – must have started forming at about the same time. It has therefore become customary to refer to the average age derived from meteorites as 'the age of the solar system', and to suppose that this also provides a measure of the age of the Earth since its formation.

The disintegration of uranium also produces some other information that can be used for dating purposes. In the process of breaking down into lead, uranium produces helium gas as a by-product. The longer uranium has been decaying, the more helium accumulates in the rock. So we could set up a radioactive clock by comparing the amounts of helium and uranium present just as readily as by using lead and uranium. You would obviously expect the two methods to come up with the same age, since it is the same uranium which is disintegrating in both cases. Sometimes this is what you do find, but in other cases the two ages derived are considerably different. The clue to why this should be so comes when you compare the two sets of ages for a range of meteorites. Where the ages are discordant, the uranium/lead age agrees with the average meteorite age we have quoted before, but the uranium/helium age is younger. This means that the meteorites concerned must have lost some of their helium. Now uranium and lead are both solids at the temperatures common on Earth or in the asteroid belt, so when lead replaces the radioactive uranium it stays put where it is. But helium is a gas, so when it is heated it may diffuse out of a rock and disappear. The occurrence of discordant ages therefore makes sense if some meteorites have been heated since their formation. Indeed, it should be possible to work out when the heating occurred by looking at how much helium has been generated by uranium since the event.

In terms of gas production, the most interesting process is not the breakdown of uranium, but the disintegration of radioactive potassium which leads to the formation of the gas, argon. Potassium is a very widely distributed element in terrestrial rocks and meteorites, so it provides a widely-used radioactive clock. Argon, like helium, is a chemically unreactive gas. It will sit around in a rock unless it is heated, when it will seep out. Consequently, potassium/argon ages can be affected in the same way as uranium/helium ages. We dated the Barwell meteorite by the potassium/argon method, and came up with an age of approaching 4500 million years. This showed that the Barwell meteorite had lost little argon, and therefore had not been strongly heated during its past history. But several meteorites give younger ages when dated in this way.

In one case, several meteorites that came to Earth at different times and places proved to have been heated at about the same time – four million years ago. How can we explain this? More generally, how can meteorites be reheated at a later stage in their careers? There is one possible heating process that could solve this problem: when two bodies hit each other at high speed quite a lot of heat is generated. If collisions have occurred in the solar system, the discordant ages can be explained readily in this way. Where several meteorites lost their gas content at the same time, this would simply imply that they were all involved in the same collision. The question of how common collisions are in the solar system is looked at in the next chapter.

5
Death to Dinosaurs?

The first astronomical object I ever saw through a telescope was the Moon, and in this I was certainly typical of many budding astronomers. I remember finding the dark patches on the Moon rather dull at first glance, but being fascinated by the bright parts of the Moon. Here, even with a quite small telescope, I could see myriads of craters: the bright regions were totally peppered with holes.

This first glance at the Moon was in the days just before the start of the space age. (I began my career as a research student in astronomy on 1 October, 1957: three days before the first artificial satellite lifted into orbit.) At the time, there was a major dispute going on concerning the origin of lunar craters. Some thought that craters on the Moon were caused – like most craters on Earth – by forces acting from below which led to eruptions through the surface. Others saw them as caused by bodies coming from outside the Moon, which dug holes of varying sizes when they hit the lunar surface. This is one of the disputes in astronomy that space exploration has resolved satisfactorily. Close-up photographs of the lunar surface and subsequent landings on the Moon by US astronauts have shown fairly conclusively that most lunar craters were produced by the second mechanism – impact.

When we discussed the fall of the Barwell meteorite in Chapter 2, we saw that it was slowed down greatly by the Earth's atmosphere. By the time it hit the ground it had lost almost all the speed it had in space. Consequently, the meteorite fragments only dug a short distance – a maximum of 1 ft (30 cm) – into the ground, making holes of about the same width as the fragment. The Moon has no atmosphere, so fragments of the same size falling on the Moon would not be slowed down at all and would hit the lunar surface at a speed of a few miles per second. Hardly surprisingly, this makes a great difference to what happens on impact. Stopping a body moving at this kind of speed produces tremendous heat and pressure: the effects are very similar to an explosion on the surface. Bodies the size of the Barwell fragments would blast out craters many yards across on the Moon, evaporating most of their own substance away in the process.

We should feel grateful that the Earth's atmosphere protects us from such happenings. But this is not the whole story. Our atmosphere only provides a very thin veil round the planet. If you could drive a car as easily vertically as you can horizontally, it would take only two or three hours to be out in space along with the satellites. Consequently, though the atmosphere can protect us from small lumps of rock, it can do little to stop the progress of larger objects. If a meteorite the size of a large house hit the Earth's atmosphere, the residue (after much melting) would still be travelling at high speed when it hit the surface. It would produce an impact crater on the Earth much like those seen on the Moon.

This is not just theoretical speculation. There is a famous crater on Earth, the Meteor Crater in Arizona, which was formed by a body of this sort of size. (It ought really to have been called the *Meteorite* Crater, of course, but it received its name earlier in this century, when the distinction between meteors and meteorites was still rather blurred.) The crater is three quarters of a mile (1.2 km) across and several hundred feet deep and looks similar to most lunar craters. Like them, it has a flat bottom dug out below the height of the countryside around; also like them, it is relatively circular. Because impacting bodies from space move so rapidly, the angle at which they approach the surface makes little difference, so the explosion they produce almost always leads to fairly circular craters. Some volcanic craters can be distinctly non-circular by comparison. Where the Meteor Crater differs from lunar craters is in not having a high rim of debris round the edge of the crater. The reason why this should be so is obvious when we consider the age of the crater (determined by the methods described in the preceding chapter). It formed as the result of an impact which occurred 25,000 years ago; there has been ample time for wind and water to have eroded away material that originally stood round the crater edge. So, despite the slightly different appearances, we can confidently equate the formation of large craters on Earth with those on the Moon.

It is abundantly clear that the Arizona crater was caused by an impact, since fragments of the original impacting body still survive. Scattered round the rim of the crater are many small pieces of iron meteorite thrown out at the impact, while below the crater floor the rocks have been shattered, compressed and heated in a most unusual way for terrestrial rocks. The Arizona crater is very young in terms of the Earth's age. If we look round for other large terrestrial impact craters, the odds are that they will be a good deal older than the Arizona crater. This means that they will be more eroded, and any

25 OVERLEAF *The Meteor Crater in Arizona should be called the 'Meteorite Crater'. It was formed about 25,000 years ago by the impact of a large iron meteorite, and is about three quarters of a mile (over a kilometre) across. Most of the original rim round the crater has now been eroded away.*

fragments of the original impacting body will have long since disappeared. Apart from the circular shape, the only evidence for an impact origin will be the peculiar jumble of rocks below the crater floor. The Arizona crater provides an excellent indication of what to look for, and, using this as the criterion, people have suggested nearly a hundred additional sites of impact craters on Earth. This is a much lower number than can be seen on the Moon, but, when rapid erosion and tectonic activity on Earth are taken into account, it actually implies a terrestrial rate of crater production which is not too different from that estimated for the Moon.

Space probes to Mercury and Mars have shown that these planets also have cratered surfaces. Mercury has no atmosphere, and its surface looks very much like the Moon's. Mars has a very thin atmosphere, much more tenuous than the Earth's, and its surface is between the Earth's and the Moon's in terms of number of craters. Venus has a much thicker atmosphere than the Earth and is so cloud-covered that its surface can only be surveyed by radar (which readily penetrates cloud). Even so, the radar maps suggest that this deep atmosphere has been insufficient to keep out the biggest impacting bodies, because they show some large circular features which are generally interpreted as being craters. The major planets have atmospheres so deep that we cannot measure down to any surface, but space probes to Jupiter and Saturn have revealed craters on the surfaces of their satellites.

The vast majority of the craters we see on all these bodies have evidently been formed by impact. The implication is that, in times past, there has been a flow of bodies circulating across the planetary orbits from Mercury out at least to Saturn. These objects have been sufficiently numerous to pepper all the planets and satellites with their collisions. The problem is that the number of craters you can count on the Moon seems much greater than you would expect from the number of Earth-crossing bodies available today. So the first question we have to ask is – when did the cratering take place?

We can get some line on the relative ages of lunar craters simply by looking at them. The older craters are often partly overlaid by younger ones and even on the airless Moon some erosion of craters occurs. (Though erosion on the Moon is certainly a very slow process: the footprints left in the dusty lunar surface by the astronauts will still probably be visible a million years from now. Maybe it is a pleasant thought for them that, if human history fails, they at least have made their mark.) These differences can be used to grade lunar craters into older and younger categories, a process which shows that the older craters are considerably more common. But this does not tell us how old the 'older' craters actually are. A clue to the actual age of craters on the Moon has been provided by dating

26 *A footprint left behind on the surface of the Moon by an Apollo astronaut. With any luck, this may last for a million years or more before it is eroded away.*

returned lunar rocks by radioactive methods. From the information obtained from this material, it appears that much of the lunar surface was formed a very long time ago. Most of the craters are over a thousand million years old, and the older craters more like three times this age.

If we look at other solar-system bodies, we find that the Moon's cratering record is fairly typical. There was an increased collision rate, and therefore many more small bodies circulating between the planets, in the earlier days of the solar system. Unfortunately, it is difficult to tell whether there was a pulse of collisions between 3000 and 4000 million years ago, with fewer collisions both before and after this time, or whether there have been two phases in the history of collisions, the first occurring from the origin of the solar system down to about 3000 million years ago when collisions were frequent, the second a less violent period from 3000 million years ago to the present day. The reason for this uncertainty is that the bombardment has been so severe that the earliest craters may have been totally obliterated by subsequent impacts. Our simple division into younger and older impacts is probably as good as any.

Looking at the period of bombardment covering the last 3000 million years, how does the rate of crater production over this time compare with the known number of planet-crossing bodies today: in other words, with the number of long- and short-period comets and asteroids? From the observations of these different groups and their orbits, we can estimate the likelihood of one colliding with a terrestrial planet (or the Moon). Apart from Mars, which is more likely to be hit by an asteroid because of its nearness to the asteroid belt, there is an approximately equal chance that a planet will be struck by a comet or an asteroid. Very roughly, the predicted rate of impact per unit area of surface is much the same for all the terrestrial planets. For craters a few miles across or more, the rate of production is about one per million square miles (2.5 million km^2) per million years. This means, for the Earth, that one would expect about three such craters to have appeared on the continents over the past million years and this fits the observed record with almost suspicious accuracy. There are three known craters which fit the bill – two in the USSR and one in Ghana.

Cratering predictions for the satellites of the major planets are much less assured. Asteroids and short-period comets are less important in that part of the solar system, so we would expect current bombardment to be lower than in the inner solar system. Moreover, there is a problem in trying to discern the effects of the early bombardment on their surfaces. The craters on these satellites are less stable than those on a body like the Moon, partly because the surface materials of several of the satellites include ice, which is a much more malleable material than rock, and so allows craters to fade more readily. Whether for this or some other reason, various satellites of Jupiter and Saturn have fewer craters than we would expect on their surfaces.

Apart from what they tell us about the rate of bombardment in the past, craters also provide information on the sizes of the impacting bodies. The size of crater that an impact will produce obviously depends on the size of the incoming body, but it will also depend on the speed at which the body hits the surface. This speed will vary from planet to planet, since they are at different distances from the Sun. The closer a body is to the Sun, the more rapidly it moves; so the inner planets are bombarded more vigorously than the outer planets. But for any given planet, the relative size of a crater is fairly simply related to the size of the impacting body. So, the number of craters of different sizes on a planetary surface can be used to work out the distribution of sizes amongst the incoming bodies. In the case of the Moon, which is the best-examined example, craters range in size

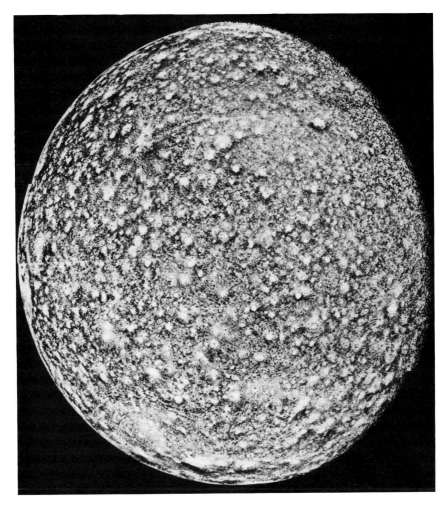

27 Callisto is the fifth satellite of Jupiter (counting outwards from the planet). This composite space probe picture shows that its surface is peppered with impact craters. They do not look like craters on the Moon (see Fig. 30), because the nature of the surface is different.

from nearly 200 miles (300 km) across to the microscopic examples that were found on the rocks brought back from the Moon. In terms of size distribution, these craters illustrate very well the astronomical rule that the smaller things are, the more frequently they occur. This distribution holds for both the younger and the older craters. The main difference observable from Earth between the two groups comes at the larger end of the size range. The dark patches on the Moon's surface are far larger than anything now referred to as a lunar crater, but they, too, were almost certainly caused by impact. Bodies of a size large enough to form that kind of hole fortunately no longer appear to be circulating through the inner solar system.

The next problem is to try and relate the size distribution derived from craters with that of small objects in the solar system today. Since our knowledge of cometary nuclei is limited, the comparison is best made with asteroids in the main belt at the upper end of the size scale, and with meteors and meteorites at the lower end. This is a satisfactory exercise as the two size distributions seem to slot together quite well. However, we are measuring the size distribution of these bodies as it is today, whereas the crater distribution gives the sizes of the impacting bodies averaged over most of the history of the solar system. This raises the question of whether it is possible to show that there has been no change in the size distribution of these small solar-system bodies over this great period of time.

One of the things that many people, including myself, have looked at over the years is the effect of collisions on the sizes of asteroids in the main belt. A quick calculation based on the number of asteroids in the belt and their relative speeds shows that quite a large number of collisions must have occurred since the beginning of the solar system (assuming, from the meteorite evidence, that the asteroid belt has been there all this time). As we saw in Chapter 3, there is also observational evidence for fragmentation in the belt. Intuitively, you would expect fragmentation to change the size distribution of the asteroids as time passed and that the size distribution today would differ from that derived from ancient craters. But the two seem to be similar. How do we solve this conundrum?

Part of the answer lies in the process of fragmentation. If you take a block of concrete and wallop it with a hammer, it will tend to break up into a few large fragments and a larger number of smaller ones, and this is what is likely to happen in all kinds of fragmentation. For example, when a meteorite is broken up by air pressure as it enters the atmosphere, it, too, produces a few large, and many smaller fragments. It is quite reasonable, therefore, to suppose that the same thing happens to asteroids when they collide. But the initial size distribution already consisted of a few larger bodies and a greater number of smaller ones, so fragmentation by collision simply moves everything down the scale without altering the overall distribution. The only difference is that there will be fewer large objects at one end of the scale, and a greater number of very small fragments at the other end. In terms of cratering, as time passes there will be fewer very large craters, and more very small craters, but the size distribution of craters in between these limits will continue much the same throughout. This conclusion fits the observational evidence. It seems likely that the objects we see today are not greatly different from the objects around in previous eras, and are perfectly capable of explaining the appearance of past craters.

When a collision occurs in the asteroid belt, it does not always follow that both asteroids will fragment. If both are about the same size, then this is probably what will happen; but when a small body hits another much larger one, the former will fragment whilst the latter gains a crater. In either case, small fragments will be produced. It is tempting to see the origin of meteorites in these smaller fragments. If this guess is correct, then it should be possible to find some indications that meteorites have undergone high-speed collisions in space, in particular that they show evidence of the pressure and heat that would result from a high-speed impact.

Most of the minerals in meteorites formed at fairly low pressures by astronomical standards. This ties in with our deduction in the preceding chapter, based on cooling rates, that meteorites were not buried too deeply wherever they formed. However, some meteorites contain minerals that can only form at high pressures and temperatures, such as diamonds. An examination of these meteorites shows that this characteristic has nothing to do with the conditions under which the meteorite originally formed, which were normal, and that the conditions needed for the formation of these minerals were imposed suddenly during an impact. The collisional effects show up in a number of other ways, many mirroring the effects in rocks underneath terrestrial impact craters. An obvious similarity is that the meteorite is cracked and distorted by the pulse of pressure passing through it. Detailed examination of many meteorites leaves little doubt that a high proportion have been involved in collisions at some stage in their past history. This finding fits in satisfactorily with what we would expect if most meteorites derived from fragmentation of asteroids.

When I first began to study meteorites, it seemed to me that not enough attention was being paid to the effect of collisions on meteorites: I thought that more of their properties might have been influenced by impact than had been realized. Since then, much more work has been done on collisional effects, but I believe there is still more to do. For example, some properties of meteorites, such as their magnetism, may be more related to the rough handling they received in space, than to their original process of formation.

If you take an ordinary magnet and heat it up, you will find that at a certain temperature it ceases to be magnetic. Cool it down again, and at that same temperature the magnetism will reappear, the reason being that the material is cooling in the presence of the Earth's magnetism, which it partly captures for itself as soon as it can. (If you deliberately shielded the material from the Earth's magnetism whilst it was cooling, it would end up non-magnetic.) A careful examination of terrestrial rocks shows that several of these have become magnetized in this way. They were originally hot – perhaps part of a molten flow of lava – and acquired some of the Earth's magnetism as they cooled. If meteorites are examined using the same techniques, many of them are found to be slightly magnetic, raising the question of whether they acquired their magnetism by a method analogous to terrestrial rocks.

This question proves surprisingly difficult to answer. To see why, we must divert to consider how planets become magnetic. Planetary magnetism tends to leak away as time passes; so, if a planet is to remain magnetic, it must be capable of regenerating its magnetism. The generally accepted mechanism through which regeneration is thought to occur is called the *dynamo process*, because it operates much like a dynamo on a bicycle or car: mechanical effort (e.g. by pedalling the bicycle) is turned into electricity by the dynamo. What is not generally realized is that you cannot generate electricity without also producing magnetic effects. (You can detect these by bringing a compass up close to a dynamo.) How does the process of regeneration apply to a planet? First of all, there must be a region within the planet that conducts electricity – like the wires in a dynamo – and, secondly, there must be some way of keeping this conducting region in motion. In the case of the Earth, the central regions are believed to consist of molten iron, which is known to be a good conductor of electricity. This hot liquid is kept sloshing about, partly because it is boiling, and partly because the whole Earth is spinning round on its axis. As a result of this mechanical effort, the Earth's core generates electricity by the dynamo process: this, in turn, produces magnetic effects that can be detected at the surface of the Earth. Because there are magnetized rocks on Earth dating from the earliest times, this suggests that the process has been working efficiently for most of the Earth's history.

The original question about the magnetism of meteorites can now be rephrased. Do meteorites come from the surface regions of a body (or bodies) which produced magnetism by the dynamo process? But this form of the question raises a serious difficulty. Small bodies have difficulty in generating electricity and magnetism by the dynamo process because any central conducting region must be very small. As a result, it seems unlikely that adequate motions can be generated within such a confined space. It is probably necessary to have a body the size of the Moon in order to obtain a conducting core that is sufficiently large to generate electricity, and even that may be an underestimate. It is not absolutely sure from lunar exploration that even the Moon managed to produce its own magnetism.

Most evidence, as we have seen, suggests that meteorites come from small bodies, but the presence of magnetism – if generated by the dynamo process – suggests they come from larger bodies. This apparent contradiction might be resolved by going back to my point concerning the unforeseen consequences of impacts. Some people, including me, have speculated that the heat and pressure of a collision can affect the magnetic properties of the colliding bodies. The more radical approach is to suppose that magnetism can actually be generated by the collision, but the approach I looked at some years ago is less radical. We have seen that the solar wind carries with it some of the Sun's magnetism and this means that interplanetary space, including the asteroid belt, is magnetic. When two bodies collide, it is possible for this interplanetary magnetism to be compressed and for the bodies to become magnetized. The process should have worked even better in the early history of the solar system, for there is some evidence that both the solar wind and interplanetary magnetism were stronger then. This has been suggested, for example, by observations of stars that are like the Sun, but not so old. Although the idea that meteorites can be magnetized by impact is still speculative, some support for it can be drawn from experiments in the laboratory. At any rate, it is possible to conclude that meteorite magnetism does not necessarily imply the formation of meteorites in large bodies.

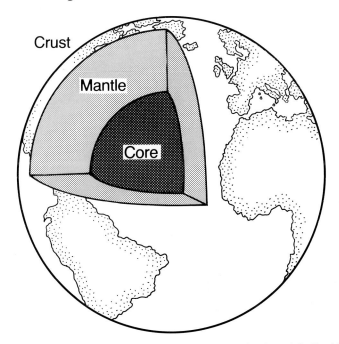

28 The Earth's interior is made up of a central core, which is mainly liquid, surrounded by a solid mantle. The latter is covered by a relatively thin veneer of crustal rocks. The interiors of other terrestrial planets are probably basically similar, but do not always include a liquid core.

To produce magnetism by the dynamo process, the body concerned has to be spinning on its axis. All the planets spin to some extent, but the larger planets tend to spin faster than the smaller ones. The Earth rotates faster than any of the other terrestrial planets, but a day on Jupiter is less than half the length of a terrestrial day. The asteroids go against this trend: despite their small size, many spin round two to three times as quickly as the Earth. Once again, collisional effects may have produced this difference. Two bodies colliding are as likely to hit at a glancing angle as head-on, and when this happens the effect is similar to what happens if you play a slicing shot at snooker. Starting with two bodies that are not spinning at all, you end up with two rapidly spinning bodies. Consequently, the distribution of spin rates in the asteroid belt depends on the history of collisions between the fragments. Moreover, whereas snooker balls are spherical, most asteroids are not; this, too, has an effect on rotation, because the rotation of non-spherical bodies is usually more affected by collisions. So the shape distribution of asteroids is clearly important.

Asteroids are never more than points of light when seen through a telescope, so it might seem a rather hopeless task to try and determine their shapes. In fact, however, the situation is not so bleak. We can see asteroids because their surfaces reflect sunlight. If the asteroid is a smooth sphere, like a snooker ball, the amount of light it reflects will always be the same, even if the asteroid is spinning round on its axis. But in the case of an asteroid which is cigar-shaped (something like the 'elliptical billiard balls' mentioned in Gilbert and Sullivan's *Mikado*), the area reflecting sunlight is less when the cigar is end-on to the observer than when it is side-on. If the cigar is rotating, its brightness will fluctuate, and this will be true even though the body can only be seen as a point of light. Turning this argument on its head, if the light reaching us from an asteroid is varying in a regular way this means, firstly, that the asteroid is not spherical and, secondly, that it is spinning at a rate which corresponds to the period of the light variations. Given sufficient observations, it is possible to obtain a reasonably good idea both of the shape of an asteroid and of the length of its day.

We have been doing a fair amount of work along these lines at Leicester. Obtaining the necessary observations is a lengthy and rather tedious business, but it is well worthwhile for the information that can be acquired covers more than just shape and spin period. A spherical snooker ball only reflects the same amount of light as it spins round because it is painted the same all over. If one face is painted white and the other black, the amount of reflected light will vary as the ball spins. An observer will see regular light variations even though the body is spherical. Fortunately, it is easy to separate

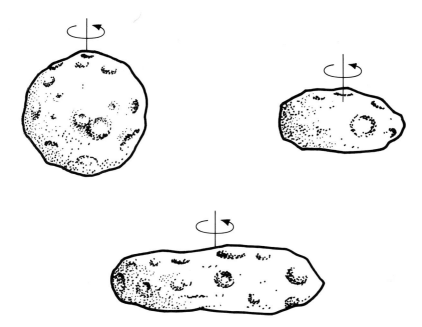

29 *Asteroids vary in shape from the nearly spherical to the highly elongated. When a non-spherical asteroid spins on its axis, the amount of sunlight reflected to an observer on Earth changes with time. Observation of these changes makes it possible to determine the asteroid's spin rate and the position of its spin axis.*

this type of light fluctuation from the variation brought about by shape. If shape is causing the change, the variation will remain the same whatever colour band you use for measurement. If the fluctuations are caused by differing materials on the asteroid surface, the amount of variation will change for measurements made in different colours.

The front face of the Moon, for example, has several large dark patches on it, whilst the face away from the Earth does not. If the Moon were to spin round every few hours like an asteroid (instead of always keeping the same face pointing towards us), this difference in the surface would produce a colour variation. So colour measurements of asteroids can be used to tell us something about the nature of their surfaces, even though we cannot see these directly. This information can be supplemented by looking for small, rapid changes in brightness as the asteroid rotates. These may be due to the appearance of large surface features (such as mountains, or craters) coming round the edge of the asteroid. When this happens, there will be slight changes in the amount of sunlight reflected and these tell us more about the asteroid surface.

Fitting together these pieces of information like a jigsaw gradually builds up a picture of what an asteroid looks like. Many asteroids

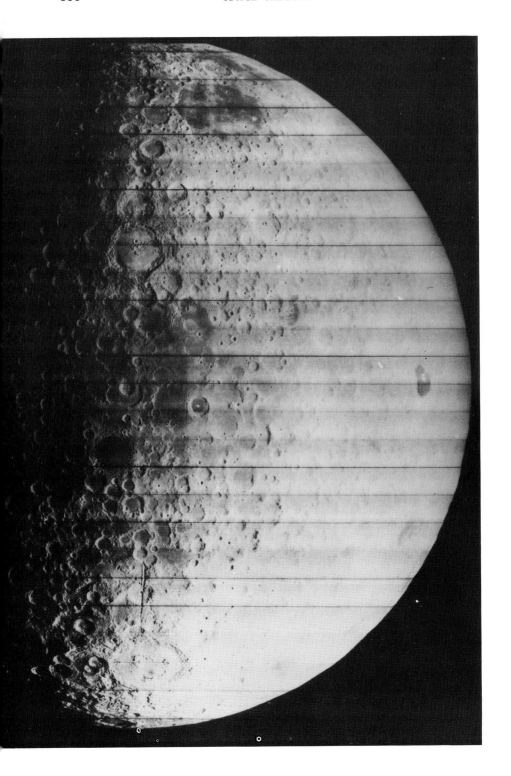

30 *The front face of the Moon (above) is not entirely covered by impact craters: large areas are taken up by lava-filled basins called 'seas'. The photograph on the left, taken by a satellite orbiting the Moon, shows part of the Moon's rear face, which is always hidden from terrestrial observers. It is evident that this face has few lava-filled basins.*

seem to resemble the moons of Mars, which is hardly surprising since the Martian satellites, as we have seen, may well be captured asteroids. However, the important point about the satellites is that they have not only been observed from Earth (like the asteroids), but have also been examined in close-up by space probes to Mars. So if the Martian moons can be linked with asteroids, then they can give us some direct information about asteroid surfaces.

Both satellites are non-spherical: Phobos is about 17 × 13 × 12 miles (27 × 21 × 19 km) in size, whilst Deimos measures 9 × 8 × 7 miles (15 × 13 × 11 km). Deimos has noticeable colour variations across its surface, whereas there are only small changes on Phobos. Both are heavily cratered, but the largest crater – over 6 miles (9 km)

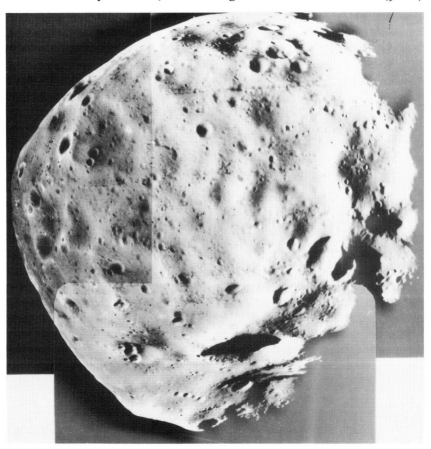

31 The Viking Orbiter 1 satellite obtained the close-up pictures of Phobos, the inner satellite of Mars, which have been used to form this composite picture. The part of Phobos to the right and bottom of the picture is in shadow. The small satellite is elongated and covered with impact craters. It is thought the surfaces of small asteroids would look very much like this.

across – is on Phobos and is large enough to alter the apparent shape of the satellite. If the body that made this crater had been slightly larger, it would probably have split Phobos in two. The craters are hemispherical holes with slightly raised rims, quite similar to lunar craters. The most interesting characteristic of the surfaces of these moons is the widespread presence of a thick layer of dust. A layer of this sort is a well-known feature of the Moon's surface, where it is referred to as the *regolith*. There it has been built up from the many small rock fragments produced by collisions with the lunar surface. It is not surprising that similar fragments are generated by collisions on the satellites' surfaces, but what is interesting is that the material stays there. A small body, only a few miles across, has very little gravitational pull and so loses material very easily. Most of the fragments produced in a collision would be expected to disappear into space. Consequently, although the larger asteroids would be expected to have regoliths, the amount retained on bodies 10 miles (15 km) across will be fairly small. The fact that this is not true of the Martian satellites may be because of the influence of Mars nearby; if so, this is one point where we cannot extrapolate from the satellites to the asteroids.

One intriguing point concerns the densities of the Martian moons. Though they look like balls of solid rock, they are estimated to be less than twice as dense as water. This is far below the density of terrestrial rocks, or even of meteorites, so it raises the question of what the moons are composed. It is quite likely that the density estimates, which are difficult to make, are wrong, but some astronomers do expect to find low-density asteroids because of what may happen when a collision takes place. If the two colliding bodies were moving relatively slowly, it is possible that a number of the fragments might stay together, rather than separate. In that case, they might look like a solid body to a distant observer, though they actually contained appreciable empty space. This would make the apparent density of the body less than the density of the individual fragments. This is an intriguing picture, though it is not clear how stable such a cloud of fragments would be after the passage of time. Moreover, close-up pictures of the Martian satellites show no sign of them being extensively fragmented throughout, so the density problem remains.

Just as the satellites revolve round Mars, so a smaller fragment could move round a larger one to form a *double asteroid*, and there may be some tentative evidence for this. One method of measuring the sizes of asteroids is to watch their progress across the sky against the background stars, when it happens occasionally that an asteroid actually passes in front of a star. The length of time for which the starlight is cut off gives a measure of the size of the asteroid. It has been claimed that, in a few instances, the starlight has been extinguished twice. One disappearance corresponded to the known asteroid, the other, it has been argued, was due to the presence of a smaller unknown companion.

Unfortunately, the evidence for double asteroids is not very strong. However, there is one result from space exploration which supports the idea of fragments moving together. When the astronauts visited the Moon they left behind on the surface instruments to measure moonquakes (tremors – like earthquakes, but much weaker – which occur in the outer layers of the Moon). The instruments were also able to record tremors due to the impact of meteorites on the lunar surface. It was found that most of the impacts were single, but, very occasionally, a swarm of meteorites would strike the Moon's surface over a short period of time. We have seen that the dust from comets moves round the Sun in streams, so producing meteor showers on Earth. The occurrence of meteorite swarms might be thought to link them in a similar way to comets, but there seems no reason why groups of meteorites should not be generated by collisions between asteroids.

A large collision with the Earth, or the Moon, is about equally likely to be due to either an asteroidal fragment or a cometary nucleus. The Meteor Crater in Arizona was almost certainly caused by an asteroidal fragment. We also have evidence for an event that fits what we would expect as a result of a cometary collision. Because much of the material in a cometary nucleus evaporates very easily, it is likely that there would be rather less debris lying about after such a collision than after an asteroidal fragment had hit the Earth. Similarly, since some of the material is highly chemically reactive, we might expect that there would be an explosive interaction with the atmosphere as the comet came down. The event that fits these requirements occurred near the Tunguska River in Siberia in 1908. A huge explosion, apparently before the descending object reached the Earth's surface, shook the atmosphere from North America on one side to the UK on the other. People up to 150 miles (250 km) away were blown over by the blast. The explosion released large quantities of dust into the Earth's upper atmosphere, the reflected sunlight from which made nights brighter for several weeks.

Scientific expeditions to the region of the explosion were delayed for twenty years, because World War I and the Russian Revolution intervened. When investigators eventually reached the area, they found no sign of a major crater, nor any obvious meteoritic fragments. It has therefore been proposed that the Tunguska event was caused by a comet. As the explosion would have been much larger if an entire nucleus had been involved, a small chip off a cometary nucleus is thought to be responsible. Indeed, some

32 A tektite from Australia. This australite shows clear evidence of having been strongly heated during passage through the atmosphere. Part of the original glass sphere can still be seen on the back (top) of the tektite. The front of the sphere (bottom) became molten, and was pressed back by the air pressure to form the flange. The ripples produced in the molten glass by the air flow can still be seen.

astronomers have tried to identify the fragment as having come from a well known short-period comet called Comet Encke.

Cometary impact may also have been involved in creating one of the odder sort of objects found on Earth. Small glassy fragments, perhaps an inch (2.5 cm) or so across, are found strewn over a few regions of the Earth's surface varying in size from a country to a continent. These objects – called *tektites* – can be picked up, for example, over wide areas of southern Australia. As a result the members of this group are usually referred to as *australites* to distinguish them from other tektites. I once took a handful of these tektites to an extra-mural astronomy course I was running, and challenged the members to tell me what they were. After a slight pause, one replied 'fossilized dinosaur droppings', a description that certainly reflects the unusual appearance of tektites even if it wasn't quite accurate. They seem to have started life as balls of molten glass, the size of big marbles, which solidified quickly. During their subsequent career the balls must have moved through the Earth's atmosphere at high speed, since some show the same sort of melting as meteorites. The origin of tektites is still a matter of controversy. The most obvious explanation envisages an explosive impact, perhaps by a comet, on the Earth's surface which fused the surface rocks into glass, and blasted the fragments out into space. The

AN
ALLARM
TO
EUROPE:

By a Late Prodigious

COMET

ſeen November and December, 1680.

With a Predictive Diſcourſe. Together with ſome preceding and
ſome ſucceeding Cauſes of its ſad **Effects** to the *East* and
North Eaſtern parts of the World.

Namely, *ENGLAND*, *SCOTLAND*, *IRELAND*, *FRANCE*, *SPAIN*,
HOLLAND, *GERMANT*, *ITALT*, and many other places.

By *John Hill* Phyſitian and Aſtrologer.

The Form of the *COMET* with its Blaze or Stream as it was ſeen *December* the 24*th.*
Anno 1680. In the Evening.

London Printed by *H. Brugis* for *William Thackery* at the Angel in Duck-Lane

rapidly-chilled glass froze in space; then, as the tektites fell back to Earth, friction with the upper atmosphere remelted their surfaces. At the same time, it slowed their fall down, diminishing the amount of heat. So, in the lower atmosphere, the surfaces solidified again, and the tektites assumed their final shape. Each such explosion strewed a group of tektites over the surrounding region.

Some possible sites for these impacts have been suggested, but the craters seem smaller than you would expect with such big explosions. This may be due – as with the Tunguska explosion – to the fact that most of the energy was pumped into the atmosphere, rather than into the surface. In any event, the explosions forming tektites must have been large enough to have affected considerable areas of the Earth. This leads to the question of whether there have been any impacts large enough to have affected the entire Earth.

A body does not have to be gigantic in order for the impact to have a worldwide effect: one more than 10 miles (15 km) across would probably do. It is likely that once every hundred million years or so an asteroid or cometary nucleus more than 10 miles across will hit the Earth. If it lands on a continent, it will throw up vast quantities of dust: if it lands in an ocean, then it throws up great amounts of water. In either case, the pollution of the atmosphere can change the surface temperature of the whole Earth to an extent that may produce major effects on plants and animals, wherever they are. It has been found in recent years that certain rare elements (especially iridium) occur in surprising amounts at a particular point in the geological record. The importance of this finding is that elements like iridium are much commoner in meteorites than in the Earth's surface layers, so their presence at a specific time in the past may indicate an asteroidal or cometary collision then. Since the deposits are widespread, we can speculate that this impact corresponded to one of our global explosions. The interesting point is that this collision occurred 65 million years ago, at a time when many plants and animals, including the fabulous dinosaurs, died out and disappeared from the fossil record. Perhaps, as many scientists think, it is all coincidence; but, if not, the rise of the mammals, including human beings, owes a passing debt to space garbage.

33 Cometary collisions – the final word.

"Still, the Jurassic period was all right while it lasted"

6
Unusual Garbage

So far we have been looking in general at the types and habits of space garbage, but in this chapter two objects of particular interest will be looked at in more detail – one asteroid and one comet. Neither object is typical, but it is the odd objects that often tell us most.

The asteroid I have chosen is Vesta. At 350 miles (550 km) across, it is one of the largest asteroids, and was discovered soon after Ceres at the beginning of the nineteenth century. It can be seen easily through a small telescope, not only because of its size, but also because it is one of the most highly reflecting of all asteroids, reflecting a quarter of all the sunlight falling on its surface. Vesta orbits the Sun at a distance of about 2.4 AU in the inner part of the main belt. Its path does not depart much from a circle, and is not greatly inclined to the ecliptic. In other words, there is nothing especially distinctive about Vesta's orbit: the real interest lies in its physical characteristics.

Vesta is whitish in colour. This, together with its great ability to reflect light, distinguishes it from other asteroids, and leads to its classification as a U-type. There are no other asteroids known with identical properties. To examine the nature of its surface in more detail requires good spectra. Fortunately, since Vesta is so bright, this raises no problem: its spectrum has been examined in more detail than that of almost any other asteroid. There are deep depressions in the spectrum of the reflected sunlight which can be identified as being caused by two different kinds of silicate minerals – the measurements are now sufficiently precise that it is even possible to estimate the relative amounts of these present.

Vesta's spectrum is very similar to spectra produced in the laboratory by a small group of meteorites. These are a sort of achondrite, and are usually referred to as *eucritic* meteorites. The correspondence between the asteroid and eucritic-meteorite spectra – both of which are unusual – raises the hope that in this instance it is possible to link directly laboratory and telescopic work. We will suppose for the moment that eucritic meteorites really are fragments from Vesta's surface, and explore what information they can give us about Vesta.

The meteorites are odd, in part, because they have clearly solidified from molten rock, whereas this is certainly not true of most stone meteorites. Using radioactive dating methods, it is possible to estimate that it took the meteoritic material a maximum of 200 million years to cool down and solidify; it also appears that the cooling process started at the time assigned to the origin of the solar system (4500 million years ago). So the surface layers of Vesta were molten, at most, for the first 5 per cent of the lifetime of the solar system and they have been solid since. The meteorites also have fewer of the lighter elements than you would expect in terms of the cosmic abundance scale, or even in comparison with the Moon's surface. As the ease with which elements boil off from molten rock depends on the gravitational pull of the parent body, the scarcity of light elements implies that the meteorites formed on a body smaller than the Moon.

The minerals in the meteorites formed at quite low pressures, but have been subjected to higher pressures since. These pressures were evidently caused by impact, since the eucritic meteorites have been pulverized into fragments which have then been re-welded together, a typical effect of impact. Also the dating of individual fragments within a single meteorite makes it clear that, on average, they were welded into their present form some thousand million years after their original solidification. This picture fits in well with what we know about the lunar regolith and rocks from the pieces brought back by astronauts. The heavy early bombardment of planetary surfaces smashed the top of any lava flows present, and then pounded the resultant fragments together into rocks. If this is what happened on Vesta, the impacting bodies must have had a range of compositions – there are, for example, plenty of C- and S-type asteroids in the vicinity of Vesta's orbit. Though most of the impacting bodies will evaporate, there should be a few fragments left (as there are round the Meteor Crater on Earth), and these fragments ought then to be pounded into the surface along with the original rocks. In fact, small fragments of carbonaceous meteorites – the easiest to detect – have been found embedded in eucritic meteorites. This process of impact mixing (often called *gardening*) goes on in the presence of the solar wind, which is continually hitting the surface of the Moon, or of an asteroid. It is no surprise therefore to find that eucritic meteorites, like the lunar regolith, contain trapped solar-wind particles.

If these meteorites were chipped off the surface of Vesta, then they must have been exposed to cosmic rays during their flight from the asteroid belt to the Earth. As we have seen, cosmic rays can produce radioactive effects in meteorites and these can be used to date how long it has been since the meteorites were dug out of the asteroidal surface. The results vary from a few million years to 60 million years – relatively recently in solar system terms.

When the spectrum of Vesta is examined, it confirms that the asteroid once had a molten surface. This raises the question of whether it was only the surface layers of Vesta that were molten, or whether much of the rest of the asteroid was also fluid. We know that there was a lot of heating available in the early solar system, especially from radioactivity, and most people believe this means Vesta must have been heated inside as well as at the surface. This could lead to the asteroid separating out into at least two regions – the outer crust that we see, and an inner region made up mainly of a different kind of silicate. However, on the evidence of the eucritic meteorites, some astronomers think the separation would have extended further. The eucritic meteorites contain less iron and nickel than typical chondrites, and the metal is supposed to have been lost from them as part of the melting process. It could have gone into the underlying silicates, but it could also have drained further inwards to form a central metallic core in the asteroid. If this picture is right, Vesta is surprisingly like a miniature Moon, with a crust consisting of one sort of silicate, an inner region of another sort and possibly a small metallic core (now solid).

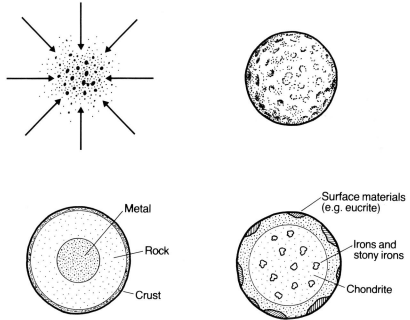

35 It is believed that asteroids formed from the coming together of large numbers of smaller particles (top). The big asteroids, such as Vesta, then probably differentiated internally (that is, the heat inside them became high enough for some of their constituent materials to separate out). The differentiation might take two forms – total (as lower left), or partial (as lower right), the latter also showing how different types of meteorite might be formed from an asteroid like Vesta, if it was broken up.

One way of testing the supposed relationship between Vesta and the meteorites is to compare their densities, but this involves determining how massive Vesta is, and that is a very difficult thing to do. Where possible, the customary way to establish the mass of an object in astronomy is to observe a natural or artificial satellite that is circling it. The orbit of the satellite will depend on the gravitational pull of the object, which is determined by the amount of material it contains. All the planets have been measured in this way, but Vesta has no natural satellite, and has never been orbited by an artificial one. The only possible way in which its mass can be estimated is to see how its gravitational pull affects other smaller asteroids in nearby orbits; but this method does not produce very accurate results, so Vesta's mass is still fairly uncertain. However, when the available data are used to work out Vesta's density, we get an interesting result: the density comes to about 3.3 times that of an equivalent volume of water. This is appreciably higher than for the other two asteroids with known densities – Ceres and Pallas – which clock in at about 2.3 and 2.6 respectively, but the interesting point is that this figure is very close to the density of eucritic meteorites, which have a measured average of 3.2. Moreover, both values are close to the average density of the Moon, so in terms of density the eucritic meteorites could certainly be surface material from Vesta, and the asteroid could also have a Moon-like internal structure.

Vesta has another feature in common with the Moon and that is that its two faces differ in their appearance. If the brightness of Vesta is monitored carefully, it is found to vary regularly, though not by very much, every five and a half hours or so. This variation depends on colour, which means – as we have seen – that it is caused by differing surface materials, rather than by shape (in fact, Vesta appears to be reasonably spherical). There are at least two types of silicate material present on the surface of Vesta and from the colour variations it now appears that they are positioned at different places on the asteroid's surface. If the eucritic meteorites derive from Vesta, this could explain why they show a fair range of properties: it would depend on the particular area from which they came. In the case of the Moon, for example, the dark patches are lava flows which have different characteristics from the cratered regions and the colours of these two areas are also different. We can speculate that Vesta seen close up might look like a miniature Moon.

A study of the various minerals in a rock can give some idea of the conditions under which the rock formed. As a molten rock cools down, the minerals in it tend to separate out in different ways depending, in part, on the gravitational pull. From the minerals present, it has been estimated that the eucritic meteorites must have formed in a body more than 10 miles (15 km) across. It is possible to estimate where that body was in the solar system by using the solar wind, some of which we know the meteorites have trapped within them. The amount of solar wind available for capturing depends on distance from the Sun (the further an object is from the Sun, the fewer the solar-wind particles hitting its surface). The amount of solar wind in eucritic meteorites indicates that they must have formed at around the distance of the asteroid belt. So the parent from which these meteorites came lies in the asteroid belt and is more than 10 miles (15 km) across. Now we believe that virtually all the bodies of this size and upwards in the main belt have already been discovered; amongst the known asteroids, only Vesta has a spectrum that matches the eucritic meteorites. It might seem then that our assumption of a relationship between the two must obviously be correct – where else could the meteorites come from?

The problem with this theory is that Vesta lies in an average sort of orbit and it is difficult to account for material being shifted from that kind of orbit towards the Earth. For this to be possible, something unusual has to happen to the orbit – like getting entangled with a Jupiter resonance, for example. There are two ways of solving this difficulty, the first of which is to suppose that our theories of how material is transported inwards from the main belt are inadequate. If this is so, then material from Vesta may be reaching the Earth, but we have not yet discovered how it manages it. Alternatively, it is possible that there used to be another body like Vesta in the asteroid belt, but that it travelled in a different orbit from Vesta, one which could readily lead to Earth-crossing fragments. This body may no longer be identifiable because it has been shattered to pieces by collisions. Most of the fragments formed in a collision will come from the interior of the asteroid, and these, as we have seen, will have different properties from the crust, so large surviving pieces from the collision will not be similar to Vesta in appearance. Some astronomers have suggested that one of the known asteroid families – called the Budrosa family – actually is this shattered Vesta-like asteroid. The observed fragments belonging to this family include some made up of the silicates expected to occur in the interior of Vesta and some of iron and nickel, corresponding to the supposed core of the asteroid. I suspect that common opinion nowadays would still point to Vesta as the most likely parent body for the eucritic meteorites, but perhaps the safest conclusion is: 'Watch this space for further developments.'

Turning from asteroids to comets, I have chosen Comet Halley to look at closely, partly because Halley is by far the most famous of comets. Moreover, Comet Halley is just returning to the centre of the solar system, and so to being visible from the Earth. This might be sufficient justification in itself for choosing it, but the main reason for talking about Halley is the same as for Vesta – it happens to be an

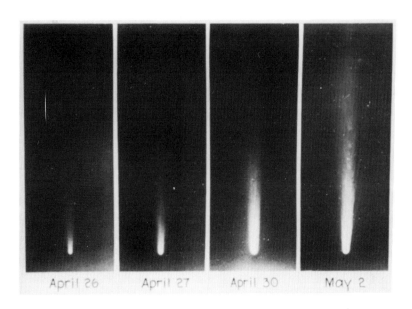

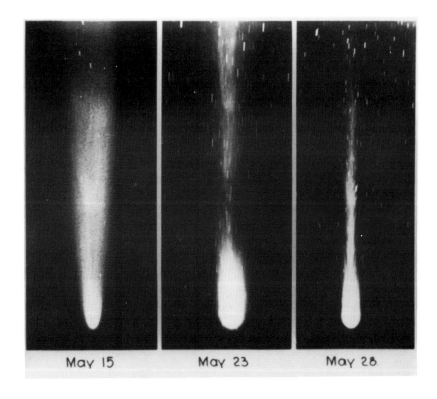

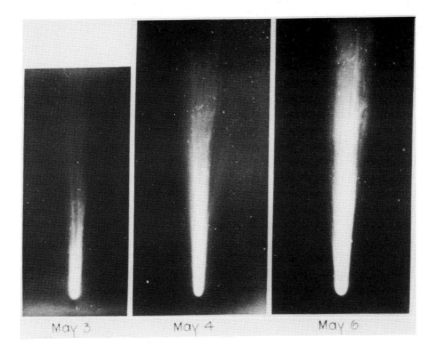

36 *This series of photographs shows how the tail of Comet Halley changed in length and appearance as the comet rounded the Sun on its last visit. The larger-scale photograph overleaf was taken on 27 April 1910. A comparison with the corresponding image in the first series shows that there can be a long, but faint tail present even when at first sight the comet appears to be quite small.*

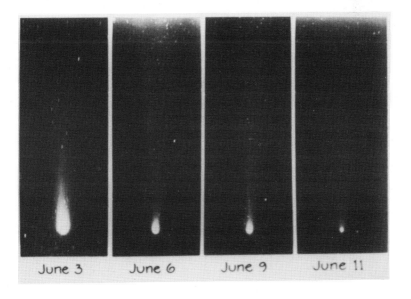

unusual object. Its most obvious peculiarity is that it has been a very bright comet for a very long time and as a result we know more about the history of Comet Halley than about the history of any other comet. As a popular rhyme (at least, popular in astronomical circles) has it:

> Of all the comets in the sky,
> There is none like Comet Halley.
> We see it with the naked eye,
> And periodically.

Halley also follows a distinctly unusual path round the Sun. As we have seen, most short-period comets have been captured by Jupiter, which means, since Jupiter is moderately close to the Sun, that they orbit the Sun once every few years. Comet Halley takes much longer – seventy-six years on the average for each trip round the Sun – and at its outermost point it reaches well beyond the orbit of Neptune. Moreover, whilst most captured comets go round the Sun in the same direction as the planets, Halley goes round in the opposite (retrograde) direction.

The orbit of Comet Halley is inclined at an angle of a bit under twenty degrees to the ecliptic. Because it has a highly elongated orbit, this means it is as much as 10 AU below (i.e. south of) the ecliptic at its furthest point from the Sun. Its point of closest approach to the Sun lies between the orbits of Mercury and Venus and there it reaches a distance of 0.2 AU above (i.e. north of) the ecliptic. The comet obviously passes through the plane of the ecliptic at two points (one going up and one going down), of which the former lies outside the orbit of Mars, the latter inside the Earth's orbit. Putting all this information together, and bearing in mind that a comet should pass close to a planet in order to be captured, poses the unsolved query of how Comet Halley managed to get captured in the first place.

The further an object is from the Sun, the more slowly it moves. At its greatest distance from the Sun (*aphelion*), Halley is moving at less than a mile (1.5 km) per second (round about the speed of the best jet fighters on Earth), a point that it last reached in 1948. Since then it has been gathering speed as it swoops in towards the Sun, and it will reach its maximum speed of nearly 35 miles (55 km) per second early in 1986, as it passes its closest point to the Sun (*perihelion*). This variation in speed means that the comet spends most of its life in the outer parts of the solar system. It only comes within Jupiter's orbit for about fifteen months of its 76-year cycle and, since comets only become reasonably bright at closer distances than Jupiter, this means that Halley is only easily visible for a small fraction of the time.

Although the period for one trip round the Sun averages out at 76

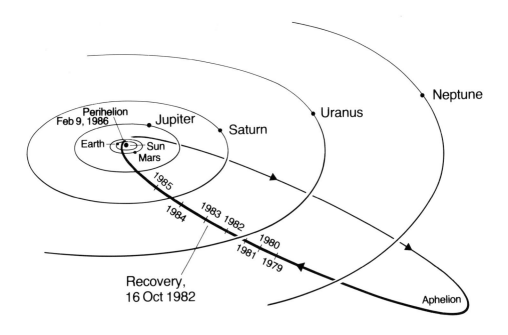

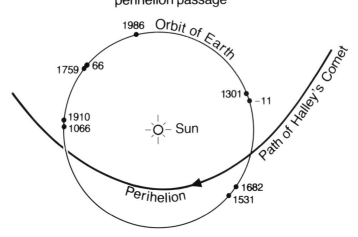

37 As the first diagram shows, the orbit of Comet Halley is appreciably inclined to the orbits of the planets. The comet only becomes relatively easy to see through a telescope after it crosses Jupiter's orbit. How bright it can appear to us on Earth depends on the relative positions of the Earth and the comet as the latter passes round the Sun. These relative positions for some previous appearances mentioned in the text are shown in the second diagram. The Earth's position in 1986 is not particularly favourable: so Comet Halley will only appear moderately bright.

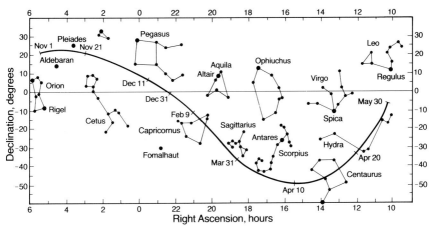

38 *Comet Halley's path through the constellations will make it visible to observers in the northern hemisphere during the latter part of 1985 and to observers in the southern hemisphere during the first part of 1986. The latter will have the better view; but the comet's passage near the Pleiades star cluster in November should provide an attractive sight.*

years, Comet Halley sometimes takes less and sometimes more time to complete an orbit. The shortest time in which it has completed the trip – 74½ years – happened last time round, between 1835 and 1910, while the longest time for an orbit – over 79 years – occurred 1500 years ago. This time round (1910 to 1986) it is doing a good average orbit. The reason for these differences is that every time the comet sweeps round the Sun, it has to pass by the orbits of the major planets. Since it travels some distance below their paths, the gravitational pulls never become too large, but they are still sufficient to alter slightly the way the comet moves, and so change the length of time each orbit takes. The amount of the pull obviously depends on where each planet is as Comet Halley passes, and this is why the period involved varies erratically on either side of the average.

As the comet brightens for a fairly short time, how well it is observed at each appearance depends on the Earth's position. The relative positions of the Earth and Comet Halley for the next appearance in 1985–6 are fairly average – the comet has been both easier and more difficult to observe at different encounters in the past. The closer the comet is to the Sun, the brighter it is, so it is best to observe it as close to that time as possible. However, it is also necessary to observe it at night time, which means it must be reasonably well separated from the Sun in the sky. The comet will be visible through a typical pair of binoculars from about November 1985 to June 1986. It will be briefly visible to the unaided eye for a short time after Christmas 1985, disappearing into the glare of the Sun in January, but it should be easier to see when it reappears on the

other side of the Sun early in March 1986. You may then be able to
follow it with the unaided eye for as much as a couple of months. This
may seem a good length of time, and so it is, but whether you can take
advantage of it depends on where you live. At this appearance, the
part of the orbit before the comet passes round the Sun is better seen
from the northern hemisphere, but once it has been round the Sun,
the comet is better seen from the southern hemisphere. Therefore, if
you want to see Comet Halley to best advantage, you should book a
trip to somewhere like Australia.

It is worth trying to see Comet Halley with the naked eye, if only
because it makes you realize how closely our ancestors must have
watched the sky before the telescope was invented. Every return of
the comet has been well observed since 87 BC, and there is now evidence that Babylonian astronomers saw it even earlier. Because
comets were long regarded as omens (usually, but not always, of evil),
some appearances of Comet Halley have figured prominently in
history. Some have tried to identify the return in 12 BC with the Star
of Bethlehem which guided the Wise Men, and the Italian painter,
Giotto di Bondone, who saw Halley at its return in 1301, included it
as the Star in his fresco of the Adoration of the Magi. However, even
allowing for the uncertainty in the dating of Christ's birth, this date
seems too early. The appearance of Halley in AD 66 also had religious
overtones, as it was held to have announced the destruction of
Jerusalem. Skipping a thousand years, we come to the most famous
appearance of the comet in European history. The death of King
Harold at the Battle of Hastings in 1066 was widely held to have been
foretold by the return of the comet in that year and it was included in
the Bayeux Tapestry woven some years later to commemorate the
Norman victory.

Gradually, scientific interest began to assert itself. Observations of
the return in 1531, for example, were used by Apian to show that
comet tails point away from the Sun. In 1682 the comet was observed
by Halley himself, but the crucial event that has made this Halley's
comet came twenty years later. Halley, after lengthy delving in old
records, managed to prove that the comet was periodic: it returned at
regular intervals to the vicinity of the Sun. On this basis, he predicted
that it would return again in 1759. It is difficult for us today to realize
what an impact this prediction had on Halley's contemporaries.
Comets had always been regarded as essentially unpredictable (and
so had been held in awe even by the well-educated) and therefore the

39 This Punch *cartoon from Comet Halley's last appearance in 1910 reminds us
how much technology has advanced in the years between. We now have a vastly
wider range of instrumentation to observe the comet than could have been imagined
in astronomers' wildest dreams at the beginning of the century.*

PUNCH, OR THE LONDON CHARIVARI—May 25, 1910.

THE GREAT AMATEUR.

Aviator. "MARVELLOUS FLIER! AND DOES IT FOR LOVE!"

return of Halley's comet in 1759, as he had forecast, played a valuable part in establishing modern science.

Another return worth mentioning is the last one in 1910, as this was the first time Comet Halley had appeared since the development of photography. It is therefore the only appearance for which we have really detailed records, comets being much better depicted by photographs than by the human eye. The observations from 1910 are being extensively analyzed now to act as a guide to the current return. Photographic techniques have improved greatly since 1910 and applying modern techniques to the old pictures has brought out details unseen by the astronomers who made the original observations. In this way, it has been possible to show that the nucleus of the comet emitted jets of material as it passed round the Sun. This finding can be related back to the comet's orbit. The comet does not move exactly as expected, even after full allowance has been made for the gravitational effects of the planets. The anomalies have been put down to the jet effect resulting from the ejection of material, which slightly shifts the direction in which the comet is moving. The jets found on the 1910 photographs confirm the possibility of this effect. Since the ejected material is seen to be spiralling out from the nucleus, these photographs also show that the nucleus is spinning round.

Dust from Halley can be observed even when the comet itself cannot as there are two meteor streams associated with it, which cross the Earth's orbit at different points. One corresponds to the region of the ecliptic which the comet passes through going upwards (from south to north), and the other to its passage through the ecliptic downwards. The first meteor shower is called the Orionids (the meteors appear to shoot out of the constellation Orion), and can be seen every October. The second shower is the Aquarids (appearing from the constellation Aquarius), and is visible in May.

For this return of Comet Halley, astronomers are mounting one of the biggest observing campaigns ever attempted. Observations will be made with every type of instrument available: ultraviolet, visible, infrared and radio radiation from the comet will be monitored by telescopes both on Earth and in space. At the peak of the campaign, it will be observed from all round the world by hundreds of professional astronomers and many more amateurs, and it is expected that the data collected will form the basis against which all

40 Modern methods of analysis can even extract new information from old photographs. This series of images obtained in 1910 (to the right) have been enhanced recently to show the presence of complicated jets shooting out from the centre of the comet (to the left). Measurements of these jets have made it possible to estimate the spin rate of the nucleus.

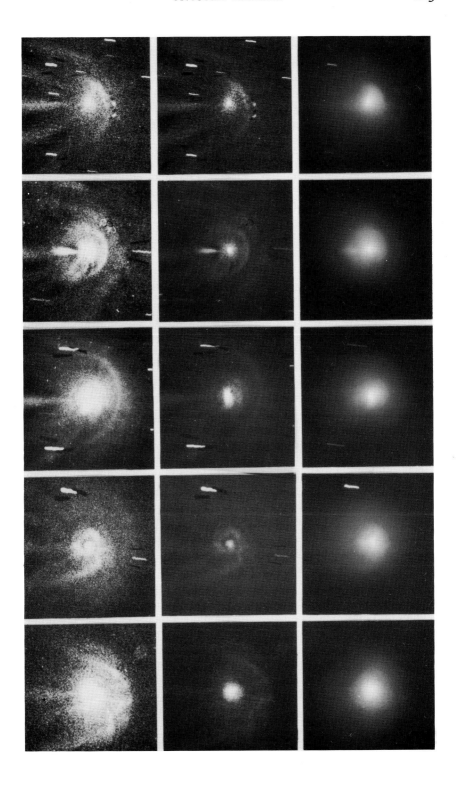

other cometary observations will be measured. An international body, called the International Halley Watch (IHW), has been set up to coordinate this programme; it aims to provide information (e.g. on the current position and brightness of Comet Halley) both to astronomers and to the public. One of its most important activities is to make sure that the vast amount of data coming out of the campaign is properly stored and made available. The much smaller amount of data from the 1910 visit was analyzed very slowly and some of the most important material was not fully reported until twenty years after the comet's appearance (indeed, some 1910 photographs have still not been looked at in detail). The hope is that, with the aid of computers, it will be possible to handle the data from the present return very rapidly. To coordinate the UK contribution to this international effort, a national committee called the Comet Halley UK Coordinating Committee (CHUKCC) has been set up. UK astronomers have access to a number of large telescopes dotted round the world and it is obviously important that these instruments are used as efficiently as possible in observing Comet Halley. After all, space garbage rarely figures highly in the priority list for observing on large telescopes, so we must make the most of our opportunities. UK observations will be filed via a British computer network: the set up devised for this is known as CHUKIT (Comet Halley UK Information Technology) and this is linked to the IHW network. Hence the British answer to comet data-handling is – 'CHUKIT'.

There was great competition to be the first to observe Comet Halley at this return: the winners were a US group, who found the comet in October 1982 when it was 11 AU away from the Sun, and approaching the orbit of Saturn. It was extremely faint, so much so that it was generally supposed to be the bare nucleus that was being observed. On this assumption, the nucleus was estimated to be small – only 5 miles (8 km) across, or so – and reflecting some 5 per cent of the sunlight falling on it, which is a very low figure. Clearly this result would not be obtained from ice, so the nucleus must be covered in dirt. However, observations since the discovery suggest that life may not be so simple. Despite its great distance from the Sun, the brightness of the comet was found to fluctuate, so if the nucleus was what was being observed, it cannot have the same material all over its surface. In fact, observations in late 1984 and early 1985 showed that a small halo of gas and dust was already present.

The IHW is mainly concerned with ground-based observations of the comet, but Comet Halley will be the first comet to be observed in close-up by space probes. Indeed, the plans to send European, Japanese and Soviet spacecraft to intercept the comet will mean that it becomes one of the best-studied objects in the solar system. Surprisingly, the USA is not sending a space probe to Comet Halley,

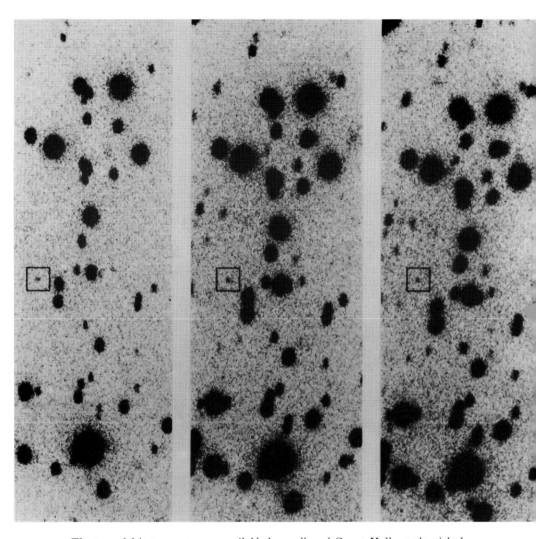

41 *The powerful instruments now available have allowed Comet Halley to be picked up far earlier on its inward leg to the Sun than ever before. These pictures, taken with the 200-inch (5-m) telescope at Palomar Observatory, date to the beginning of 1984, and show the comet at a distance of 800 million miles (1300 million km) from the Sun. The comet (in the box) is a good deal fainter than the background stars, but can be distinguished by its motion relative to them.*

but to make up for this they are sending one to a different comet, called Giacobini-Zinner. The spacecraft they are sending on this trip has a fascinating history. It started life in 1978 as ISEE-3 (the third in a series of satellites labelled *International Sun-Earth Explorer*), and was intended to measure the interaction between the solar wind and the Earth's magnetosphere. From 1982 onwards, ISEE-3 was

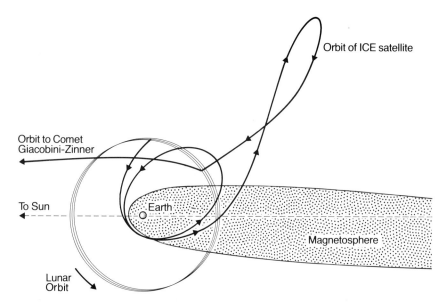

Orbit of ICE satellite

Orbit to Comet
Giacobini-Zinner

To Sun

Earth

Magnetosphere

Lunar
Orbit

42 The ISEE-3 (Third International Sun-Earth Explorer) satellite was originally launched to study the magnetotail of the Earth. It was realized, however, that, if it could be manoeuvred to make a series of close passages to the Moon, then the lunar gravitational pull would dispatch the satellite on an orbit towards Comet Giacobini-Zinner. The manoeuvres were successfully carried out, and the satellite – now renamed ICE (International Cometary Explorer) – was on its way at the end of 1983 for an encounter with the comet in September 1985.

gradually shifted out of its original orbit by means of successive encounters with the Moon. A very close lunar fly-by at the end of 1983 allowed the gravitational pull of the Moon to redirect the spacecraft into an entirely new orbit. This will intersect the orbit of Comet Giacobini-Zinner in September 1985, several months ahead of the planned encounters with Comet Halley; at this intersection the US spacecraft is expected to pass through the tail of Giacobini-Zinner at a distance of anything from 3000 miles (5000 km) to ten times that distance from the head. Measurements on this comet and Halley should complement each other, since Giacobini-Zinner is a less dusty comet than Halley.

It takes extra rocket power to lift a space probe out of the plane of the ecliptic, so the encounters with Halley are planned to occur as the comet passes through the ecliptic. It also turns out that more rocket power is required to reach the first crossing point in November 1985 than the second in March 1986, so all four spacecraft have been targeted to arrive at the later date, as Halley passes down (from north to south) through the ecliptic. The Japanese probe will pass furthest away from the nucleus at a distance of about 65,000 miles (100,000 km). The two Soviet spacecraft are arriving at Halley via Venus (where they are each due to dump a planetary probe before continuing on their way). The first should pass through the comet at a distance of over 6000 miles (10,000 km) from the nucleus; if it experiences no problems, the second probe will be allowed to pass closer, perhaps at 2500 miles (4000 km) from the nucleus. Plans for the European probe are the most ambitious of all. It is intended to pass at only 300 miles (500 km) or so from the nucleus.

The European probe is called Giotto after the Italian who painted the comet in the fourteenth century. Its extensive instrumentation is intended to analyze the gas and the dust in detail, as well as to establish the existence and properties of the nucleus. The investigators looking after the various instruments come from several West European countries, including the UK. Because it is passing through the central regions of the comet, dust will be a particular hazard to the European spacecraft. Although the particles may be quite small, Halley is travelling in a retrograde path round the Sun; this means that dust will impact on the spacecraft at a very high speed, and so be able to cause considerable damage. This is why the USSR decided to test the cometary environment with one space probe before committing another. The European decision to send their probe through the densest part of the dust cloud is therefore a calculated gamble, but even if the dust does score major hits on the spacecraft, it should be possible to gather enough data to justify the exercise. The information from all the probes will be combined with the ground-based observations to provide the most detailed view of any piece of space garbage we have yet had.

7
Life, the Universe and Everything

Many of the questions raised in previous chapters have been deferred to this one and the reason is that there is no certain answer to them. So although I will be providing some answers here, what I have to say should be taken with a large pinch of salt. Indeed, I really ought to start my account with one of the traditional fairy story beginnings, such as, 'In the High and Far-Off Times. . . .'

The most fundamental question that needs to be tackled in this chapter is how the solar system started, for the origins of asteroids and comets must somehow be tied in with the answer. Because this is one of the central questions of all astronomy, very many solutions have been suggested down the years; but most of these concentrate on explaining the presence of the Sun and the planets, and the small bodies of the solar system tend to drop through the cracks and be forgotten. However, it is particularly in the study of origins that space garbage becomes important; after all, quite a lot of it has been around, not greatly changed, from the start. But I shall begin by sketching out the main lines of observational evidence for the formation of stars and planets.

It is useful to begin by standing back from the solar system in order to see how it fits into the larger environment. In Chapter 1 we considered the view from a nearby star, but, for a good view of the Sun's surroundings, we must go further out to (say) hundreds of times the distance to nearby stars. From this distance we would, of course, only see the stars: planets would be invisible. You would find that the Sun is just one of a vast assemblage of stars – perhaps a hundred thousand million of them – which exist as a group well separated in space from other similar assemblages elsewhere in the Universe. Our own assemblage is referred to – a little parochially perhaps – as our Galaxy. Despite the huge number of stars involved, there is a very definite structure in the way they come together, so that our Galaxy has a highly characteristic shape when viewed from a distance. The stars are mainly restricted to a volume of space which is much like a discus in shape, or maybe like a pair of cymbals held together. There is a large bulge of stars in the centre surrounded by a disc of stars which thins out to a moderately well-defined edge. Within this volume, the stars in the disc are moving in a regular way round the central bulge, just as, in the solar system, the planets circle the Sun. The Sun itself occupies a fairly undistinguished position in the disc about two thirds of the way out from the centre towards the edge. At this distance, it takes the Sun 250 million years to complete one orbit round the Galaxy.

Besides stars, our panoramic view of the Galaxy would also show clouds of gas and dust, usually described simply as *interstellar matter*, dotted about at intervals in the disc. Some of the clouds are so hot that much of their dust has evaporated and the remaining gas is glowing with the heat; others are extremely cold. The main factor determining the temperature of a cloud is whether there are any hot stars in the vicinity of each cloud and, if so, how many there are. But, regardless of temperature, all the clouds are strung out in fairly well-defined lines through the disc. In fact, if you concentrate on the cloud patterns, you can see that they spiral out in a reasonably regular way from the central bulge, defining what are called the *spiral arms* of our Galaxy. Imagine, if you can, putting an octopus in a bowl of water, holding it by its head and spinning it round rapidly. The way in which the arms stream behind the spinning head would be very similar to the way the arms of the Galaxy stream out from the central bulge. (At least, I hope so: I must admit I have not tried the experiment.)

The existence of clouds of interstellar matter has been known for over half a century, but it is only in the last ten years or so that the existence of giant clouds, much more dense than the average, has been fully realized. The reason for calling them *giant* clouds becomes obvious as soon as you work out the amount of material they contain. One of them might have a hundred thousand or even a million times the amount of material contained in the Sun. In the central parts of such a cloud, the density of material may be ten thousand times or more what it is for ordinary interstellar matter. (Even so, this still corresponds to vacuum conditions in a terrestrial laboratory: interstellar space is very empty.)

In the last few years, a lot of effort has been put into developing instruments to detect microwave radiation from space. Microwaves lie between heat (infrared) and radio radiation, and have some of the characteristics of both of these. They are best known for producing heat in microwave ovens: the microwaves leaking out of these ovens can affect nearby radio telescopes – to the annoyance of radio astronomers. A lot of work is being put into microwave instruments in astronomy because they can provide otherwise unobtainable information about interstellar gas. Over the past few years, microwave telescopes have detected a remarkable range of substances mixed into the gas, ranging from items well known on Earth, such as water, through less frequently encountered substances, such as

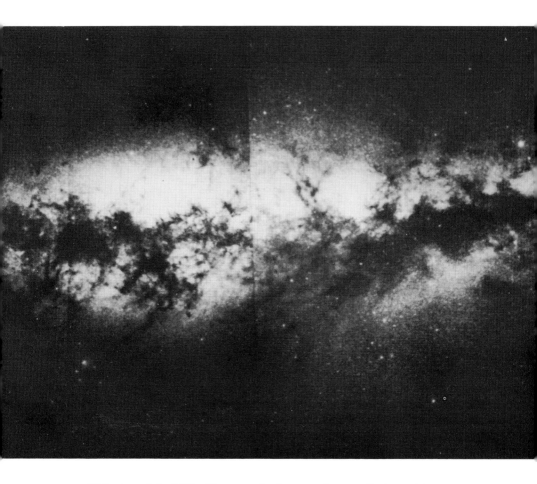

43 *This view of the Milky Way stretches between the constellations of Sagittarius and Cassiopeia. The white regions are the myriads of stars in our Galaxy. The dark patches are clouds of dust blocking the light from the stars behind them.*

poisonous hydrogen cyanide, to the totally unfamiliar. An example in the last category is a fairly complex chemical containing mainly carbon, but with a little hydrogen and nitrogen. Chemists have given it the almost unpronounceable name (for astronomers) of cyano-decapentayne. The important point about this wide range of materials is that they are not distributed uniformly throughout interstellar matter, but are actually concentrated towards the centres of giant clouds.

The central regions of the giant clouds have also been examined with infrared telescopes, which give slightly different information from the microwave detectors and show that the giant clouds have low-temperature stars embedded in them. 'Low temperature' is, of course, a relative term. A star like the Sun has a surface temperature

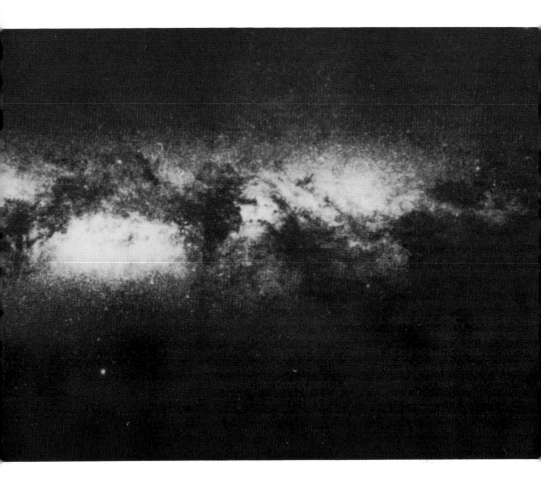

of around 6000° C (11,000° F), while the stars we are talking about, although still hot by terrestrial standards, may be at a third of this temperature, or less. The important point is that they are at a lower temperature than you would expect for an ordinary star.

These observations are important for what they imply about star formation. It has been believed by astronomers for at least a couple of centuries that stars form out of interstellar matter. The process involves the material in a cloud falling inwards under its own gravitational pull. As the pull of the material in the central parts of the cloud attracts the outer parts, and draws them inwards, the material at the centre becomes increasingly compressed as a result. Since this pull inwards obviously exists for all clouds, the first question to answer is why all interstellar clouds have not long since collapsed. The main reason is the existence of heat. A hot gas tries to expand and the gravitational pull inwards in a hot cloud cannot usually win against the tendency to expand outwards. As we have seen, interstellar clouds will be hot if they are warmed by nearby hot stars. For a gas to be cold, so that it can condense, it must be shielded from these stars. The easiest way of doing this is to surround an inner region of the cloud by a thick outer layer of cloud, so that the latter then absorbs all the stellar heat, leaving the central part cold and able to contract. This is simply a description of a giant cloud. Microwave observations not only detect what substances are present: they also indicate the cloud's temperature. The central regions of a giant cloud can be very cold, perhaps 200° C (400° F) below the freezing point of water. As stars begin to condense in these regions, they produce heat, so increasing the temperature of the nearby cloud. This can be detected by infrared observations.

In one sense, the formation of stars is a self-limiting activity. As the condensing star (usually called a *protostar*) contracts to its final state, it gradually heats up until it reaches a proper stellar temperature. At this point it starts heating the surrounding material left over from its own formation, and blows it away, thus making sure that no other stars subsequently form too close to it. However, as we have seen, even the central regions of a giant cloud are far more massive than an individual star and what happens – as infrared observations confirm – is that a whole group of stars condense together. The group then blows away all the nearby interstellar matter, and we are left with a star cluster. The stars in the cluster may stay together for a long time – perhaps a thousand million years – but more typically, as the gas disappears, the stars wander off on their separate paths round the centre of the Galaxy. As the surrounding material disperses, the young stars become visible through an ordinary optical telescope.

The most famous giant cloud near us is the Orion nebula. The old pictures of the constellation Orion show it as a hunter with a shield in one hand and a club in the other. It is perhaps unintentional that, in terms of stars, he is one of the most empty-headed characters in the sky. To balance this, the three stars that make up Orion's belt are easy to see. Below them the dagger (or sword) which hangs from the belt is marked by a fuzzy blob: this is the Orion nebula. If you examine the nebula in the infrared, you can distinguish a number of stars currently forming; some have developed sufficiently to blow away surrounding interstellar material, and can be seen and photographed through an ordinary small telescope.

The observations give a satisfactory degree of agreement with our expectations regarding the origin of stars. At first sight, there seems

44 The photograph shows the constellation of Orion. The large number of faint stars present, and the different sensitivities of the camera and the human eye to colour, make it difficult to trace the normal shape of the constellation as seen in the sky, so this is shown in the accompanying diagram. The Orion nebula can be seen as a hazy blur some distance below Orion's belt.

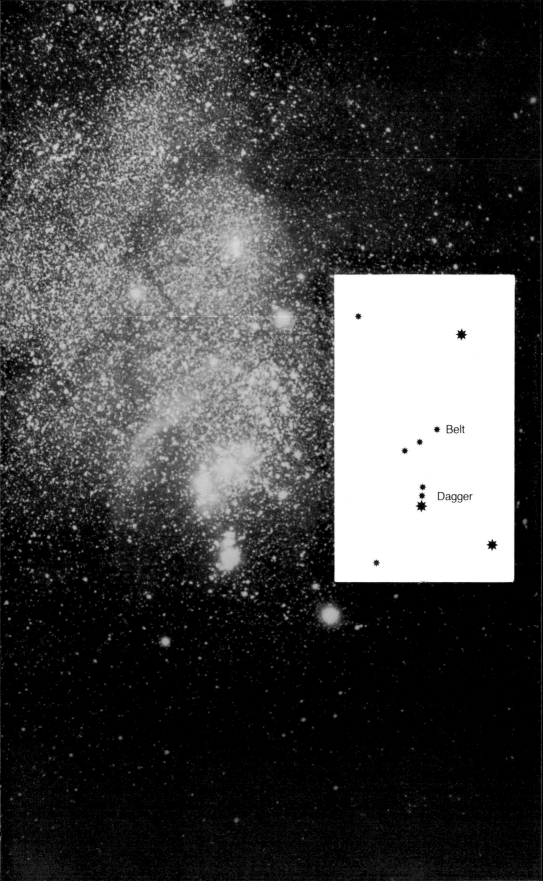

Belt

Dagger

little hope of observing planets in a similar way as they form, but here, too, giant clouds can provide some insight. Microwave data tell us that gas in the immediate vicinity of protostars is flowing away from the stars, but not to an equal extent in all directions. Instead, it shoots out from the star in two jets going in exactly opposite directions. The interpretation of this finding seems to be that the star has a disc of gas and dust round its equator and that this disc prevents material from leaving the equator, channelling it instead into jets from the north and south poles of the star.

This observation can be linked back to the solar system. When you rotate a body, it tends to flatten out if it can. Thus, our Galaxy is flattened because it is spinning round, and the same is true of the material round protostars. But we know that the solar system is also flattened – to the ecliptic – with the planets moving round the Sun in the same plane and the same direction, just like the material in the disc moves round a protostar. So is it possible that planets form from such a disc? Most astronomers – though by no means all – would answer that they do.

This possibility could resolve one of the basic problems the solar system offers – the division between the terrestrial and the major planets. The gas discs round protostars may contain some exotic materials, but their overall composition is certainly close to the cosmic abundance scale, so any planets formed from such discs might at first sight be expected to have the same composition as the central star. But this overlooks the heating effect of the young star, which must give the material in the inner disc a much higher temperature than material in the outer disc. Heavier materials – such as iron/nickel and silicates – can condense quite happily at moderately high temperatures, but gases are much more difficult to pull together in hot regions because they will try to expand, rather than condense. Since the two lightest gases in existence, hydrogen and helium, expand very readily, a planet could only contain these substances if it formed from a part of the disc at some distance from the central star. Moreover, while the inner parts of the disc are struggling to retain their gases against the increasing heat, escaping material from the star begins to sweep through the disc. We saw that the jets of gas from the protostar initially escape from the poles, but as the star reaches its final stages of formation, the stellar wind strengthens still further and, blowing out now in all directions, sweeps away loose material in the disc.

Putting these developments together, it is clear that the inner parts of the disc will lose any substances which cannot condense rapidly into solids at moderate temperatures. The outer parts of the disc, on the other hand, will be cooler and less affected by the stellar wind, so they should be able to retain most of the elements present. If planets condense from this disc, the inner ones will consist primarily of iron/nickel and silicates, whilst the outer ones will be mainly hydrogen and helium. This is obviously not too bad a description of our own planetary system.

The way I have described it, the formation of planets sounds a pretty commonplace event. This has been a key belief in a controversy that has raged over the past twenty years or more. The argument concerns the question – how common is life in our Galaxy? If there are large numbers of planets, then the chance of life occurring elsewhere is obviously much higher. When I started work in astronomy, one of the arguments for planets being common ran as follows. Many stars spin round rapidly, but the Sun spins quite slowly, and the reason for this was said to be that the Sun has shared some of its rotation with the planetary system. The argument then was that rapidly-spinning stars have no planetary systems, whereas slowly-spinning stars do have planetary companions. As there are a vast number of slowly-spinning stars in the Galaxy, this view concludes that planets, and therefore opportunities for life, must be frequent.

In recent years, the viewpoint has changed. An examination of young stars which are expected to age like the Sun shows that they are actually spinning fairly quickly. They are similar to the Sun in being magnetic bodies with a wind streaming out from them, but they differ in the fact that the stellar winds are much stronger than our own solar wind. This difference is now seen to be crucial, for it has been shown that a rapid spin rate can be slowed by the loss of material. As the wind weakens, so its effect on spin rate decreases. It is therefore thought that the young stars will come to spin more slowly until, when they reach the Sun's age, they will spin at much the same rate as the Sun. If this idea is correct, then the Sun's rate of rotation has nothing to do with the fact it has planets.

Recent measurements by IRAS (the Infrared Astronomical Satellite) have opened up the question of other planetary systems in a different direction. This satellite has made it possible to look in some detail at the amount of solid material circulating round other stars. This is not the same as detecting planets: the infrared satellite preferentially detects what we have called space garbage. However, we must bear in mind the English saying 'After the Lord Mayor's show comes the muckcart'. The equivalent here is that after the planets comes the space garbage, so the presence of the latter may hint at the presence of the former. For example, one of the brightest stars in the sky, Vega, has been found to have sizeable quantities of space garbage circulating round it. Vega is in the constellation Lyra (the Harp), and is very easily visible from the northern hemisphere. I have a soft spot for it because my first research on stars, carried out during a period in California, concentrated on an area of sky close to it and I always used it as a rough guide to the region I wanted. Now the significant point about Vega is that it has rubble lying around it despite the fact that it is a rapidly-spinning star. This association therefore not only supports the view that spin rate and the occurrence of planetary material are not connected, but also suggests that stars much hotter than the Sun might have planets.

It is probably fair to include observations of stars of all ages in talking about the number of planetary systems in existence. But can observations of present-day protostars be happily extrapolated backwards for 5000 million years to the formation of the solar system? In fact, where we can cross-check this extrapolation, it does seem to work. For example, we have seen that meteorites can trap solar-wind particles, so an examination of meteorites which have not been strongly heated since their formation allows us to analyze such particles from the early days of the solar system. They appear to occur in greater abundance than we would expect from conditions today, and this fits with our observations that younger stars today have stronger winds.

We have seen that some meteorites reflect conditions very shortly after the formation of the solar system: in particular, the most primitive – the carbonaceous chondrites – should also be able to tell us something about the actual formation processes. There have been several attempts to estimate the pressure and temperature conditions under which this type of meteorite formed. All are rough, but they do indicate that the discs observed round protostars should provide an adequate starting point for forming such material. According to the evidence we have discussed before, this means that it is possible for bodies the size of asteroids to condense from such discs. In most processes where gases condense into solids (or liquids), it is the first stage of formation that is most difficult – once there is a tiny nucleus, the rest of the material condenses much more quickly. For example, when raindrops form in clouds on Earth, the tricky part is getting them started. A small raindrop will grow into a big one without too much difficulty. Correspondingly, if objects the size of asteroids can be grown, there is unlikely to be much of a problem in building these up into planetary-sized objects.

The crucial question is one of time. We have seen that the growing heat and wind from the early Sun would soon blow away any material that had not rapidly condensed into solids. The extinct radioactivity we have discussed earlier shows that asteroids (and, consequently, planets) did form quickly. Condensation of asteroids was already well under way within 10 million years of the process starting (that is within the first 2 per cent of solar-system history). So our general picture of how the solar system formed does hang together. If the bodies in the inner part of the solar system were to survive, they had to form quickly. The evidence indicates that they did, indeed, form quickly.

Actually, the extinct radioactive material in meteorites may tell us more than this. One question that we have not considered is why short-lived radioactive substances appeared in the early solar system, as there is no known way of creating them in the solar system either now, or when it was forming. In fact, current theories of how elements are created suggest that most of the radioactive substances of interest must have appeared in stellar explosions at very much higher temperatures than ever existed in the solar system. These explosions happen at rare intervals when a star becomes unstable and blows up, spreading its contents over the surrounding space. Some of the explosions are relatively small, but some are very large and these – called *supernovae* – reach such high temperatures that radioactive materials can be created and spread out by the explosion. A supernova is in fact the most catastrophic event known in our Galaxy. For a few weeks, the single star involved can give out nearly as much light as all the other stars in our Galaxy put together and, correspondingly, the exploded material that is shot out will contaminate a considerable volume of surrounding space. This is the only way we know for creating the extinct radioactive substances found in meteorites, so we must assume that when meteorites were forming – at the beginning of the solar system – the disc of material from which they came was slightly polluted by the explosion of a nearby supernova. (In fact, the range of radioactive substances in meteorites suggests that there may even have been two supernova explosions in succession.) It might seem that the chances of a supernova occurring near the early solar system were small; but, of course, if our ideas on star formation are correct, the Sun formed as one of a group of stars. So, in its early days, it had many more near neighbours than it has today. There is also the point that the material from a supernova does not simply contaminate nearby clouds of gas and dust, but actually acts like a blast from a bomb, compressing any clouds that stand in its way. It may be that what actually happened is that the Sun and its surrounding disc were contracting slowly when a nearby supernova explosion suddenly compressed them, completing the process of formation in a very short time. If this picture is accurate, then we are partly indebted to the father and mother of all atomic bombs for our origin.

45 OVERLEAF *Two photographs of a galaxy taken thirteen years apart. In the second, on the right, a supernova (arrowed) has exploded in the outermost fringes of the galaxy. The light from this single star adds an appreciable fraction to the total light from the galaxy.*

JUNE, 1959

MAY, 1972

46 A supernova in our own Galaxy. We are seeing a cloud of gas expanding rapidly outwards from the original exploding star at the centre. This is a familiar picture of the Crab Nebula: but it is clear looking at it how a supernova might compress and pollute any material lying nearby.

The evidence from meteorites is important for understanding the origin of planets: it is obviously doubly important for understanding the origin of asteroids. From the viewpoint of solar-system history, it is extremely useful that the asteroid belt lies where it does. Since it overlaps the regions of both the terrestrial and the major planets, meteorites can hopefully give us information on both types of environment. The reason why the asteroid belt occurs at all is almost certainly because Jupiter occurs where it does. The density of material in the original disc of gas and dust was highest near the centre, where the Sun formed: the inner planets nevertheless came out small, simply because most of this material was soon swept away.

Jupiter is the largest planet because it formed at the nearest point to the Sun where the light gases could survive, and therefore at the point where the amount of material left in the disc was greatest. Moreover, as the most massive planet, it had a major gravitational effect on its surroundings, even whilst it was forming, and its large gravitational pull greedily attracted any disc material in the vicinity. This did not necessarily lead to Jupiter capturing the material, but it did foil any attempt by neighbouring material to form itself into a planet. In these terms, the asteroids can be seen as the last remnants of a failed planet.

Minor planets in the asteroid belt developed as well as they could. The abundant forms of heating that were about in the early solar system had some effect on their interiors and almost certainly the iron/nickel and silicates separated out. It is still unclear whether the metal went to form a core with a shell of silicate round it, or whether small pockets of metal developed in a sphere of silicate (a bit like a fruit cake in appearance), but perhaps different minor planets developed different sorts of interior structure. If we take the differences between iron meteorites at their face value, there must have been perhaps a dozen bodies in the original asteroid belt. (Other lines of evidence suggest that this is probably about the right order of magnitude.) These 'minor planets' need to be distinguished from the present-day asteroids which are derived from them, but Ceres is almost certainly one of the original bodies which has survived. It is so large that subsequent destructive processes can hardly have affected it.

So long as the original disc contained plenty of gas, collisions between bodies would have been generally gentle, since the gas would have the same effect on bodies circling the Sun as the Earth's atmosphere has on Earth satellites. It rounded off their orbits until their *relative* speeds became quite low. In consequence, whilst the gas was there, bodies tended to stick together on collision, building themselves up into something bigger. However, once the gas was blown away, orbits in the asteroid belt began to change, and the relative speeds of the bodies in that region increased. Collisions now became more violent, leading to cratering and fragmentation. As a result, most of the original minor planets have gradually been ground down into small fragments since the days of the early solar system, so creating the asteroid belt as we know it now. This fragmentation has obviously exposed the interiors of the original bodies: if these underwent various types of internal separation in their early careers, this should be apparent now. In fact, the different sorts of asteroids that have been found suggest that they do represent fragments of an internally-separated body. This is particularly clear in the case of some asteroid families, which may consist of a mixture of different classes of asteroid. Attempts have been made to construct some idea of the original minor planet from the fragments.

The region of the asteroid belt became too hot for cometary ices to survive at an early stage in the development of the solar system, so cometary nuclei have usually been regarded as left-over material beyond the edge of the planetary system. Because the density was low there, only small bodies could form. This ties in with what we know about comets – that they appear from well beyond the furthest planet. The original estimates suggested, in fact, that cometary nuclei could extend as far as half the distance to the nearest star. Since comets may enter the planetary system from any direction, this picture implies that there is a vast shell of comets circling the Sun, but far out in space. The shell is usually referred to as the *Oort cloud*, after the Dutch astronomer who originally suggested its existence. If comets have appeared at roughly the same rate as they do now throughout the history of the solar system, the Oort cloud must contain about a hundred thousand million comets – about the same as the number of stars in our Galaxy.

In recent years, the nature and even maybe the existence of this cometary cloud has been called into question. The giant clouds we discussed earlier have been discovered since Oort put forward his suggestion of a cometary shell. As the Sun has circled the centre of the Galaxy maybe twenty times since the solar system formed, it must have passed through giant clouds on several occasions. These clouds will have little effect on the planetary system because the gravitational pull of the Sun on the planets is strong, so the system will plough through the cloud and come out the other side unscathed. The same can hardly be true of the Oort cloud. The comets are far away, where the Sun's gravitational pull on them is small, so it would seem likely that a cloud could easily detach them.

There are two possible ways in which this problem could be resolved. The more radical proposal asserts that the comets are, indeed, stripped away at each encounter with a giant cloud, but it also argues that they are replaced by new material picked up from the cloud itself. The gases detected in these clouds are very similar to the gases emitted by cometary nuclei, so the argument is that these clouds might also contain gases condensed into lumps of ice, like cometary nuclei. The temperature in the cloud is certainly low enough for this to happen. The fascinating point about this proposal is that it would mean that comets provide us with a close-up look at interstellar material, rather than solar-system material. The alternative, less radical proposal assumes that something like the Oort shell of comets does exist, but instead of being at a very great distance from the Sun, it is actually not too far beyond Pluto's orbit. At that distance the Sun's gravitational pull is still fairly large, so the cometary nuclei would be retained even during passage through a giant cloud. According to this theory, comets are once again relegated to being purely solar-system material. Both ideas raise some difficulties but I suppose the second theory is slightly preferred at present.

This controversy over the origin of comets links with another we have touched on in earlier chapters – the distinction between comets and asteroids. We have seen that there is some possibility of confusion between short-period comets and Earth-crossing asteroids, yet, unfortunately, these are just the objects we would most like to distinguish, because they seem to be the only possible sources of meteorites. There are about half a dozen objects which, in terms of their orbits, look like short-period comets, but which appear to be point objects, like asteroids, through the telescope. For example, the asteroid Hidalgo follows an orbit which is 2 AU from the Sun at its nearest point and 10 AU from the Sun at its furthest. This is much more typical of a short-period comet than an asteroid, but Hidalgo has the colour and telescopic appearance of an asteroid. Now short-period comets obviously cannot survive for ever, because every time they pass close to the Sun some of their substance evaporates away. The obvious conclusion is that they disappear entirely after what is, by solar-system standards, a short time, but it is not possible to be certain about this because you cannot be quite sure that a cometary nucleus is the same throughout. The dust we see in a comet might be small chips off a block of solid material at the centre of the icy nucleus and, if so, there will still be a small solid body moving round the Sun after all the ice has gone. The suggestion is that unusual objects, like Hidalgo, may be these 'dead' comets.

If this picture is right, then at least some of the meteorites raining down on the Earth may be from comets, rather than asteroids; on the basis of the theories about comets this might mean that we are receiving fair-sized chunks of interstellar matter each year. This is an attractive idea, but unfortunately it does not seem very likely. Apart from the large question mark over the interstellar origin of comets, there is an equally large one over the existence of solid cores in comets. In the first place, the estimated size of cometary nuclei is quite small, say 10 miles (15 km) across, so a solid core would only be a mile (1.5 km) or so in diameter. But the unusual asteroids tend to be appreciably larger. Then, again, when comets have actually been seen to split up, or disappear, no solid core has ever been observed. In any case, the solid material in comets could only correspond to carbonaceous meteorites, because the temperatures and pressures required for the formation of the minerals in stones and irons are very difficult to reconcile with the interiors of icy cometary nuclei.

The question may now be resolved. In 1983 the Leicester group used IRAS to discover a new Earth-crossing asteroid, which was labelled

1983 TB. This object appeared to be an ordinary asteroid in its composition, but its orbit was most unexpected. It was following much the same path as the Geminid meteor shower and, as we have seen, meteor showers have always been supposed to come from comets. This may mean that 1983 TB *is* a dead comet, but there is another possible explanation. At the far end of its orbit, 1983 TB passes through a densely-populated part of the main asteroid belt and it could be that it experienced an impact there during a recent passage. If so, the meteors may simply be fragments that were chipped off and are now trailing behind the asteroid. From this viewpoint it could be significant that Geminid meteors seem to be denser than those from most other showers. Certainly, some recent ground-based observations we have made of 1983 TB seem to indicate that it has the sort of bare rock surface usually associated with small asteroids, in which case it is unlikely to be the core of a dead comet.

We can certainly conclude nothing final about where meteorites originated. The probability is that asteroids are the basic source, but there may be other possible origins. After all, a meteorite recently found in Antarctica has most likely reached us from the Moon.

For one important type of data it is irrelevant whether meteorites started off in asteroids or comets. The information they provide on the origin of life in the solar system is important either way. Though ideas about life in the Universe are legion, hard evidence from outside the Earth is very difficult to find. You can certainly guess at the number of planets in our Galaxy (though, as we have seen, such estimates have fluctuated violently down the years), but the only detailed evidence about life processes comes from our own solar system. It is now widely believed that life, even in its most simple conceivable form, has only developed on Earth in the solar system. On the other hand, what meteorites show is that the basic chemicals necessary for life can be manufactured fairly easily within a forming planetary system. The message of the meteorites seems to be that these basic materials are there: what they need is a suitable environment in which to develop. From our observations of the chemicals in giant clouds, this statement can probably be extended to our Galaxy as a whole, and indeed some organic materials in the early solar system may have survived from the original interstellar material. This implies that all planetary systems may acquire similar sets of complex chemicals when they form; but since of all the environments in the solar system only the Earth's proved suitable for life, it may be that the chemicals need something like terrestrial conditions to develop further. If this is so, life forms must always start from the same materials in similar environments and so are likely to evolve a fundamental similarity. Thus, whether planetary systems are rare or common, man may be the measure of all life.

Space Junk

When my colleagues were told the title of this book, a typical reaction – apart from distaste – was to say: 'Shouldn't it include satellite debris as well?' I, myself, tend to distinguish between space garbage, which is provided by nature, and space junk, which is man-made. However, I ought to include a few words about the second category, in case you, too, thought it should be included.

In a sense, space junk includes all artificial satellites, since they certainly clutter up space near the Earth. Communications satellites are already running into problems from over-crowding. Their favoured orbit is situated some 22,000 miles (35,000 km) above the Earth's equator, but so many satellites are being dispatched into this orbit that they could soon interfere with each other. This does not mean that they will actually collide, but that there may be difficulties in keeping their communications separate. The probability of satellites actually hitting each other is still small, though it has already occurred. A good hard collision would obviously produce fragments – just like a collision in the asteroid belt – but the satellites that have broken up so far have done so for reasons other than collision.

Perhaps the most interesting example of fragmentation was the Pageos satellite, launched in 1966. This was simply a large balloon, blown up to its full extent once it had been placed in orbit round the Earth. Pageos had a bright, highly reflecting surface. This was mainly intended for bouncing radio messages between different points on the Earth's surface, but it also meant that it could be easily seen by ground-based observers. In July 1975, the balloon broke up into twenty-eight pieces, but these were still bright enough to be tracked. Six months later it fragmented again, producing another forty-four pieces. The reason in each case can probably be traced back to the plastic from which the balloon was made. Its strength was gradually reduced by the space environment until it finally gave up. Less readily observed (and less reported) than balloon satellites are the endings of Soviet killer satellites. Their task (so far purely for testing) is to manoeuvre near to a target satellite and explode.

The closest parallel between artificial satellites and the bodies we have been talking about also comes at their death. Unless satellites orbit at a great distance, they are gradually dragged down by friction with the Earth's atmosphere. In their final stages, artificial satellites descend rapidly through the Earth's atmosphere, heating up and melting, and when this happens they flash across the sky in much the same way as a falling meteorite – indeed, they can be mistaken for one. Also like a meteorite, they may be fragmented by heat and pressure during the descent. For example, the Soviet satellite, Sputnik 4, showered pieces over Canada in 1962 in much the same way as the Barwell fireball scattered fragments in England not long after. The bigger the satellite, the more spectacular its descent and break up. Possibly the most publicized burn up in the atmosphere was that of the 75-ton (tonne) US satellite, Skylab 1, which came down over Australia and the Indian Ocean in 1979, scattering quite large fragments on the world below. Satellites are still increasing in size and number, and currently a couple of satellites decay in the Earth's atmosphere every week on average. Consequently, satellite falls are becoming as interesting as meteorite falls. Perhaps the best advice is to repeat what I said earlier in the book: 'Watch this space!'

Glossary

Cross-references are indicated in italic.

Ablation   A process in which the atmosphere heats the surface of an incoming *meteorite*, melting the surface material and blowing it away.

Achondrite   A *stone meteorite* which contains no *chondrules*.

Anti-tail   *Comets* normally have *tails* which point away from the Sun. Occasionally, a comet is also found to have an anti-tail which appears to point in a sunward direction. This is purely a perspective effect – the tail is spread out like a fan; the Earth, in this case, is so placed that we look down this fan and see the tail spread out on either side of the *coma*. To us, one side forms the tail and the other the anti-tail.

Asteroid   A small body whose size can range from several hundred miles (or kilometres) across down to a fraction of a mile. Most asteroids are believed to contain varying amounts of rock, metal and *carbonaceous* material.

Astronomical Unit   The average distance of the Earth from the Sun (approximately equal to 93 million miles, or 150 million kilometres). The Astronomical Unit, often abbreviated to AU, is used as the basic unit of length for measuring distances within the *solar system*.

Aurora   Coloured lights in the upper atmosphere, usually best seen near the north and south magnetic poles. They are caused by solar *plasma* striking the atmosphere and heating it so that it emits light. They are therefore most common when the Sun is particularly active. A distinction is sometimes drawn between the aurora borealis (the Nothern Lights) and the aurora australis in the southern hemisphere.

Binary asteroid see *Double asteroid*.

Bolide A bright *fireball*.

Carbonaceous meteorite A *meteorite* which contains an appreciable amount of material in the form of chemical compounds of the element carbon.

Chondrite A *meteorite* which contains *chondrules*.

Chondrule Many *meteorites* contain small glassy spheres: these are the chondrules. It is still not certain how the chondrules originated, though it is clear that they formed outside the meteorite and were incorporated into it afterwards.

Coma The 'head' of a *comet*, which appears through the telescope, or on a photograph, as a hazy ball of light. The coma is generally thought to be a mixture of gas and dust evaporated from the *nucleus*. Its brightness comes partly from light emitted by the gas, and partly from sunlight reflected by the dust.

Comet Usually refers to an extended object, consisting of a *coma* and a *tail*, as depicted on photographs. However, it can also refer to the *nucleus*, which most astronomers believe provides the material for the coma and tail.

Cosmic ray A sub-atomic particle moving at a very high speed. Some cosmic rays hitting the Earth are simply particles captured from the *solar wind*, but those cosmic rays travelling at the highest speeds all come from sources outside the *solar system*.

Direct motion All the planets move round the Sun in the same direction. Looking down on the *solar* *system* from the north, you would see this motion as anti-clockwise. Any body that moves in this same anti-clockwise direction is said to have direct motion. Similarly, a body spinning in this direction is said to have direct rotation.

Double asteroid   Two *asteroids* revolving round each other and held together by the gravitational force between them. Sometimes called a binary asteroid.

Dynamo process   A method of generating magnetism from the motions of liquid, electrically-conducting material in the interior of a *planet*.

Earth-crossers   Most *asteroids* move around the Sun along fairly circular paths, remaining between the orbits of Mars and Jupiter all the time. A few, however, have elongated orbits which bring them in towards the Sun, crossing the orbits of Mars and the Earth on the way. Such asteroids are called Earth-crossers.

Ecliptic   The plane of the Earth's orbit round the Sun. However, since most of the other *planets* move in much the same plane, it is often used more generally to mean the plane near which objects in the *solar system* preferentially move.

Fireball   A very bright *shooting star*. Often used as an alternative word for *bolide*, though strictly speaking a fireball is less bright than a bolide.

Galaxy   A very large number of stars held together by their mutual gravitational attraction. Our own Galaxy is a spinning disc containing perhaps a hundred thousand million stars, some concentrated in a central bulge and the remainder in *spiral arms*.

Infrared   Radiation that lies beyond the red end of the visible spectrum. It is this sort of radiation which conveys most of the heat we feel when we sit in front of a fire.

Interstellar material	Clouds of gas and dust which can be found lying between the stars in the *spiral arms* of our Galaxy.
Iron (meteorite)	A *meteorite* which consists predominantly of iron mixed with a smaller amount of nickel.
Kirkwood gaps	Regions in the *main belt*, at specified distances from the Sun, where few or no *asteroids* are found to move. These regions are called after the nineteenth-century American scientist who first noticed them, and are believed to be due to *resonance*.
Magnetosphere	The region round a *planet* within which its magnetic effect can be detected. The boundary of this region is set by the *solar wind*.
Main belt	Most *asteroids* move round the Sun in a region between the orbits of Mars and Jupiter. This general region where they are most commonly found is called the main belt.
Major planet	Describes any planet considerably larger and more massive than the Earth, and which contains appreciable quantities of hydrogen and helium.
Meteor	A small piece of matter which enters the Earth's atmosphere, and is entirely burnt up before it can reach the Earth's surface.
Meteor shower	A number of meteors which enter the Earth's atmosphere from the same direction in space (the *radiant*) at nearly the same time. Most meteor showers occur when the Earth passes through the orbit of a *comet* and encounters the dust left behind by the comet on a previous trip.
Meteorite	A lump of matter which, after entering the Earth's atmosphere from space, reaches the Earth's surface. The material is usually classified in terms of its composition as a *stone*, *iron* or *carbonaceous meteorite*.

Meteoroid A general term which covers both *meteors* and *meteorites*, whether in space, or on Earth.

Minor planet An alternative name for *asteroid*.

Nova See *Supernova*.

Nucleus A *comet* is believed to have at its core a small solid body not more than a few tens of miles (or kilometres) across. Because the *coma* and *tail* of a comet contain both gas and dust, this nucleus is supposed to be a 'dirty snowball'. The 'snow', when heated by the Sun, produces gas, and the 'dirt' provides the dust. Since nuclei are so small, it is difficult to be sure that any have been definitely observed from Earth. The 'nucleus' sometimes mentioned in descriptions of comets usually refers to a bright central region of the coma.

Oort cloud The Dutch astronomer, Oort, suggested that the *solar system* is surrounded by a vast number of cometary *nuclei* – his 'cloud'. The gravitational effect of passing stars occasionally sends one of these nuclei in towards the Sun, where we may see it as a comet.

Planet A body, at least a few thousand miles (or kilometres) across, which orbits the Sun.

Plasma A gas which has been broken up into particles with positive and negative electrical charges. The two types of particle are present in equal numbers, so the gas as a whole has no electrical charge.

Protostar A star which is still in the process of forming.

Radiant The point in the sky from which a *meteor shower* appears to approach the Earth. Individual showers are usually called after the constellation in which the radiant is situated: for example, the Perseid meteor shower radiates outwards from a point in the constellation Perseus.

Radiation belt

A region near a *planet* containing a high concentration of *plasma*. The plasma is held in place by the planet's magnetism. As a consequence, it is concentrated towards the magnetic equator of the planet (the region midway between the north and south magnetic poles), and encircles the planet like a belt.

Regolith

Continuing impacts by meteorites on the surfaces of *planets* or their *satellites* eventually break up the near-surface region into small fragments. The resultant 'soil' is called a regolith.

Remote sensing

The observation of planetary surfaces and atmospheres from instruments carried at some distance above the *planet*'s surface. Most remote sensing is carried out by *satellites*, though aircraft have been used on Earth.

Resonance

In discussions of the *solar system*, this usually refers to a gravitational interaction between two bodies which builds up over a period of time. For example, if an *asteroid* circles the Sun with a period exactly half that of Jupiter, then its point of closest approach to Jupiter occurs once every two orbits at precisely the same point in space. In consequence, the gravitational interaction between the asteroid and Jupiter builds up until it becomes sufficient to change the asteroid's orbit. This is believed to be the origin of the *Kirkwood gaps*.

Retrograde motion

The opposite of *direct motion*. Motion round the Sun in a clockwise direction as seen by an observer positioned to the north of the *solar system*. It can also refer to the direction of a *planet*'s spin.

Satellite

A smaller body moving round a larger body. In this sense, a *planet* can be said to be a satellite of the Sun. However, the term is usually reserved for bodies orbiting planets. Man-made objects are referred to as artificial satellites to distinguish them from natural satellites, or moons.

Shooting star   Refers to the bright flash of light as a *meteor* crosses the sky.

Solar system   The Sun and all the objects that are bound to it by its gravitational attraction.

Solar wind   *Plasma* ejected from the Sun which sweeps out through the *solar system*.

Spiral arm   Some *galaxies* consist of a central bulge of stars from which lines of stars and *interstellar material* trail out in a spiral pattern. These latter are the spiral arms.

Stone (meteorite)   A *meteorite* which looks rather like a terrestrial rock and, indeed, consists of similar materials. Stones are distinguished by the presence, or absence of *chondrules*. Those with chondrules are called *chondrites*: those without are *achondrites*.

Stony iron   A *meteorite* which contains some regions similar to a *stone* meteorite and some similar to an *iron* meteorite.

Supernova   An exploding star which can for a time become as bright as ten thousand million ordinary stars. A supernova is to be distinguished from a nova which, although also an exploding star, is much less bright.

Tail   The gas and dust in the *coma* of a *comet* are swept back by the pressure of sunlight and the *solar wind* to form a tail. Since the gas and dust are affected in different ways, separate gas and dust tails can sometimes be distinguished. Both point away from the Sun.

Tektite   A small body, an inch or two (a few centimetres) across at maximum. It consists

of glassy material and seems to have been
formed by impact (probably on the Earth).

Terrestrial   A planet like the Earth consisting mainly of
planet   rock and iron. It is to be contrasted with a
major planet.

Ultraviolet   Radiation that lies beyond the blue/violet end
of the visible spectrum. It is this sort of
radiation which is most likely to produce a
suntan in human beings.

Zodiacal light   A faint hazy light which stretches along the
ecliptic. It results from the reflection of
sunlight by the dust in the *solar system*. Its
name derives from the fact that the constellations of the Zodiac also lie along the
ecliptic.

Index

Figures in italic refer to illustration captions